General Preface to the Series

Recent advances in biology have made it increasingly difficult for both students and teachers to keep abreast of all the new developments in so wide-ranging a subject. The New Studies in Biology, originating from an initiative of the Institute of Biology, are published to facilitate resolution of this problem. Each text provides a synthesis of a field and gives the reader an authoritative overview of the subject without unnecessary detail.

The Studies series originated 20 years ago but its vigour has been maintained by the regular production of new editions, and the introduction of additional titles as new themes become clearly identified. It is appropriate for the New Studies in their refined format to appear at a time when the public at large has become conscious of the beneficial applications of knowledge from the whole spectrum from molecular to environmental biology. The new series is set to provide as great a boon to the new generation of students as the original series did to their fathers.

1986

Institute of Biology
20 Queensberry Place
London SW7 2DZ

Preface

Genes have come to stay as part of our essential day to day vocabulary. Whether we are breeding sheep, elaborating educational strategies, or planning public health policy, we need to know about genes and their implications for our future.

Also, although less dramatic than setting foot on the moon, the scientific breakthrough of gene manipulation is hugely important both as a human achievement and in terms of its implications for economics and medicine.

This book sets out to explain what genes are, how they are organised as molecules, and how they are regulated in cells and organisms. It is not a primer in gene manipulation; other books in this series deal very competently with that topic. Rather, it considers genes in terms of their structure and function within biological systems, information which should be mastered before the student addresses topics such as evolution, population genetics, or gene cloning.

Dr Trevor Beebee of the Department of Biochemistry University of Sussex has read and criticised all of the Chapters, and I am most grateful to him for his invaluable contribution.

Norman MacLean
1989

Contents

General Preface to the Series iii

Preface iv

1 Genes and Genomes 1
1.1 The gene defined 1.2 Gene structure 1.3 Evidence that DNA is the genetic material 1.4 The characteristics of the genetic code in DNA 1.5 Analyzing the genome

2 Small Genomes 25
2.1 Viroids 2.2 Viruses 2.3 The genomes of bacteria, their structure and regulation 2.4 Concluding remarks

3 Chromatin in Eukaryotes 52
3.1 Chromosomal proteins 3.2 The chromatin of transcriptionally active chromosomes 3.3 The mitotic chromosome 3.4 Interphase chromatin

4 Mechanisms of Gene Regulation 74
4.1 The need for regulation 4.2 How a gene works
4.3 Structure of large ribosomal RNA genes of *Xenopus laevis*
4.4 The mammalian beta globin gene 4.5 Mechanisms of gene regulation in prokaryotes and eukaryotes

5 Gene Regulation and Cell Differentiation 97
5.1 Cell differentiation and selective gene expression are correlated
5.2 Cell commitment can often be divided into distinct stages, namely determination and differentiation 5.3 Cell commitment is

normally very stable 5.4 The stability of commitment is dependent on both nucleus and cytoplasm 5.5 Commitment is often progressive in its onset 5.6 The stability of commitment is not absolute 5.7 The mechanism of commitment 5.8 The stabilisation of gene expression 5.9 Some special mechanisms of gene regulation

6 Genes in the Context of Evolution 115
6.1 Evolution at a genetic level 6.2 Mutation is a random process 6.3 DNA repair ensures that most damage to DNA does not result in mutation 6.4 The rate of change in the DNA may be quite different from that in the proteins coded by the DNA 6.5 The relationship between gene mutation and protein structure 6.6 Selfish DNA 6.7 DNA fingerprinting provides a way to determine relationships between individuals very precisely 6.8 The evolution of oncogenes

References 127

Glossary 129

Index 135

1

Genes and Genomes

1.1 The Gene Defined

Need for the word *gene* first arose out of Gregor Mendel's classical observations on peas, published in 1866, although the word was not coined until 1909 when Johannsen proposed its usage to denote an inherited factor in the genotype, following the discovery of Mendel's work in 1900 by de Vries. In order to appreciate its original significance and meaning, it is necessary to rid one's mind of subsequent genetical understanding and look again at the startling realisation of the distinction between phenotype and genotype, (see glossary for definition of these terms). By the late nineteenth century it was becoming clear in the minds of many experimentalists that the most remarkable observation concerning inheritance was that an organism could transmit to its progeny a potential for phenotypic expression not actually demonstrable in its own phenotype. Such an observation predated Mendel, since it had been clear for quite some time that parents, both possessing black hair, could produce a child with red hair. But it was not until after Mendel's period that it became evident that a genetic factor could exist unexpressed in both parents and yet be capable of transmission to, and expression in, some of their children.

This distinction between phenotype and genotype, implying that the phenotype represents only a partial expression of the total genotypic potential, still represents a foundation stone in genetics, and had created a need for a word which would designate the factor in the genotype, responsible for transmission and determination of a single phenotypic character. Such a factor came to be called a gene.

As genetics advanced, the gene was defined more precisely, especially in physical terms. Thus, when it was evident that chromosomes were the carriers of the genetic information, it became possible to define the position which an individual gene occupied on a chromosome arm. This position was designated its location, or *locus*, both within an individual chromosome, and, when the individual chromosomes could be distinguished and categorised, within the entire genome. Following this recognition that genes occupied distinct chromosomal loci, there followed the fundamental recognition of DNA as the genetic material. Although the manner of the precise arrangement of DNA in the chromosome remained obscure (and some of that obscurity persists even at the present time) it became evident that there was an approximately linear distribution of genes along a chromosome, and that the DNA formed a continuous thread running, albeit with many imposed orders of coiling and supercoiling, from one end of a chromosome to another. A gene could therefore be redefined with new precision as a length of DNA molecule, itself a sequence of nucleotides in a polynucleotide chain.

Because the gene was being actively investigated in a multi-disciplinary way in the 1940s and 1950s, it became desirable to further define the gene in terms of certain important parameters. Thus, in the 1950s Seymour Benzer proposed three further subdivisions of the basic word, the *cistron*, the *muton* and the *recon*; these were essentially operational definitions of the gene as determined or encountered by different approaches.

The cistron was defined in terms of gene function, following application of the cis/trans test, and represents the physical length of DNA that codes for a protein or RNA. A muton was defined as the smallest unit within DNA in which a change could result in mutation and it is now clear that this is in fact a single nucleotide. The third term, the recon, was defined as a unit of recombination, that is the smallest unit within the DNA capable of being independently involved in recombination, – this is now known to be the individual nucleotide. Although all of these words served useful functions in their time, they have largely fallen into disuse with increasing knowledge of DNA structure and nucleotide arrangement.

The situation now is that the word gene, although used more extensively than ever before, has been substantially overtaken by events and its use no longer carries the absolute clarity that is obviously desirable. How has this come about? Although the word gene continues to be used most frequently to designate a DNA molecule that carries in its nucleotide sequence the code for a protein or RNA, it is not absolutely restricted to this usage as we shall see in the following pages. For example, regulatory genes and pseudogenes may not necessarily subscribe to the above condition, nor is it always clear whether a gene sequence includes introns (see page 9) and even perhaps some flanking sequences which may be essential in transcription. A further area of uncertainty is the precise sequence referred to in using the word. Thus, in a population of diploid organisms, such as rabbits in a hayfield, a gene coding for the blood pigment globin might be referred to as the rabbit globin gene. But rabbits, in common with all other mammals, have multiple and differing types of globin, and so it is necessary to be more specific, referring to the rabbit beta globin gene. Since a population may have some variation due to stable polymorphism or a few mutant forms of the beta globin gene it becomes necessary either to refer broadly to the wild type, that is the most common allelic form of that gene sequence, or to actually designate a particular sequence. A further complication of this topic arises in that an individual may be heterozygous at the beta globin locus, carrying two dissimilar forms of this gene. How can it be made clear which allelic form is being designated?

It is likely that this introductory discussion has already introduced words and concepts which are unfamiliar to some readers, but hopefully all will be made clear in subsequent sections. It will be evident from this section, however, that as knowledge about genes has increased, so the precision of the term has diminished. A list of definitions of current terminology can be found in the Glossary (page 129), but some definitions are given below.

Gene

A sequence of DNA that carries the code for a protein or RNA molecule, and frequently includes regulatory regions at either or both ends. DNA sequences which are closely related by evolution or mutation to genes may retain the designation, even if no longer functional. Genes consist of DNA duplexes, only one strand of which carries the coding information. This is the sense or anti-coding strand, being effective in dictating the coding strand of messenger RNA. The other strand, the non-sense 'strand' of DNA, is not used genetically; but note that the strand acting as sense for one gene may well be in continuity with the non-sense strand of an adjacent gene. Both sense and non-sense strands together are commonly referred to as the gene. Most genes of higher animals and plants (eukaryotes) are interrupted by intron sequences, which are not represented in the messenger RNA (see discussion of introns on page 9 of this chapter). No genes in bacteria (prokaryotes) are interrupted by introns.

Structural Gene

A DNA sequence that codes for protein, thus excluding sequences such as regulatory genes. Some authors also describe genes coding for ribosomal and transfer RNA as structural genes, but more commonly the term refers purely to protein coding sequences.

Regulatory Gene (or Regulator Sequence)

A DNA sequence whose primary function is to control the rate of activity of other genes. The products of regulatory genes may be RNA or protein, or, in some usages, certain regulatory genes may have no products. Thus promoter or enhancer regions may be referred to as regulatory genes – in this book these regions will be termed promoter or enhancer 'sequences', thus avoiding such an ambiguous use of the term gene.

Pseudogene

A DNA sequence that closely resembles the sequence of a known gene at a different locus, but due to the insertion of stop signals or deletions, is unlikely to be translated or transcribed. Pseudogenes which lack introns are referred to as processed pseudogenes. It is assumed that pseudogenes have an evolutionary relationship to normal genes, and some may have arisen by reverse transcription from messenger RNA.

Many other classifications of genes exist in the literature, such as master genes, producer genes and so on, but these usually carry their own descriptions and will not be categorised here. What should be stressed is that unless the use of the word gene is clearly appropriate it is safest to refer to a stretch of DNA simply as a 'DNA sequence' or 'DNA region', and this terminology will be used elsewhere in this book. While the terminology is under discussion, it is useful to refer also to the word allele, itself a contraction for allelomorph. In theory any gene may exist in an almost infinite range of variants due to small base substitutions, but in reality, within a species or a population, the actual range is much smaller. Such variations of a gene are termed alleles of that gene, and if sufficiently distinctive, may earn the terminology of gene themselves. For example, the sickle-cell allele of the human beta globin gene is often referred to as the sickle cell gene.

1.2 Gene Structure

Although the double helical structure of DNA is very well known, and fully described in all basic texts of Biology and Biochemistry, a surprising number of possible pitfalls surround the topic, and it is as well to summarise the various physical characteristics and biomolecular properties of this remarkable molecule. These are presented below in an extended tabular form.

1.2.1 Summary of DNA structure

1. *It is a polynucleotide* of four different single nucleotides, each consisting of a purine or pyrmidine base, the pentose sugar deoxyribose, and a phosphoric acid group. The bases are adenine (A) and guanine (G), which are purines, and thymine (T) and cytosine (C) which are pyrimidines.
2. *It may exist as a single or double stranded molecule*, the latter being a double helix of two parallel single strands. When in the double helical form, the bases are connected by hydrogen bonds in specific and unique base pairing, two bonds linking the AT base pair and three bonds linking the GC base pair. Since the bases constitute the genetic code, it follows that the genetic information is actually on the *inside* of the double helix.
3. *The double helical form of DNA has strands in anti-parallel array*. This follows from the fact that each DNA molecule is polarised, one end having a phosphoryl radical on the 5′ carbon of its terminal nucleotide, the other possessing a free-OH on the 3′ carbon of the terminal nucleotide. In double helical array, one strand runs from 5′ to 3′ left to right, the complementary strand running 3′ to 5′ left to right. (DNA sequences are normally written and read running from 3′ on the left to 5′ on the right, and these may be designated as upstream and downstream respectively).
4. *The double helix comes in three alternative conformations*. These are known as the A, B and Z forms, of which B is the form most commonly found in nature. A fourth form C, is a variant of the B form. Both A and B forms are right-handed helices, the former with 11 bases per turn of the helix, the latter with 10 bases per turn. *Z* form DNA is a left-handed helix, with 12

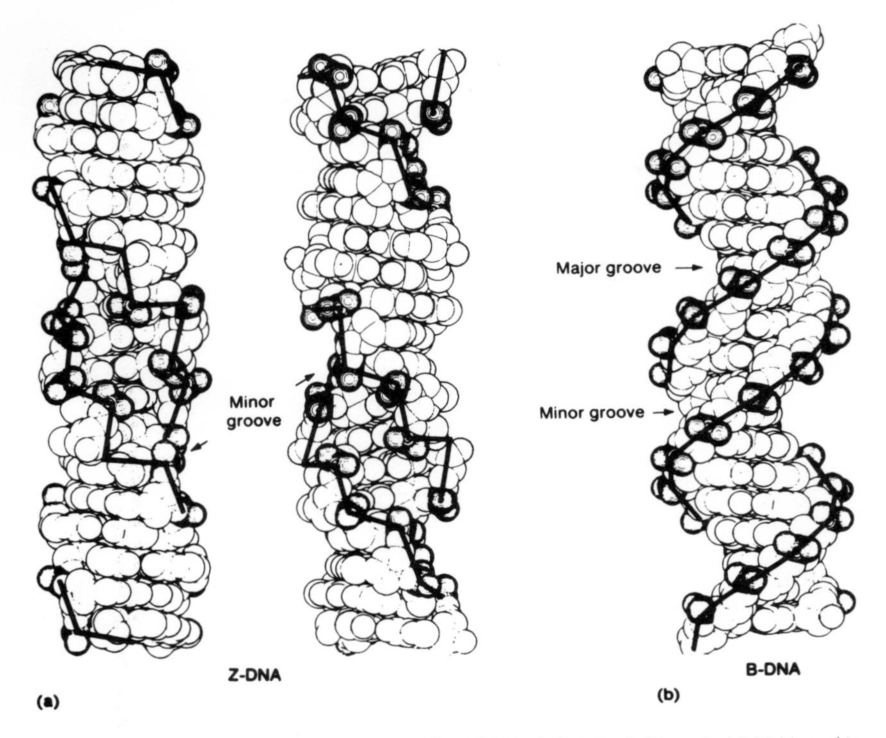

Fig. 1.1 Diagrams of space-filling models of DNA. In (a) the left handed Z DNA and in (b) the right handed B DNA. Heavy lines mark the course of the sugar-phosphate backbone, which can be seen to follow a somewhat zig-zag path in the Z form and a smooth path in the B form. (In Wang, A.H., *et al.* (1979) *Nature* **282**, 680. Kindly provided by Professor Alex Rich).

bases per turn, and adopts a zig-zag conformation. A form of DNA occurs under less hydrated conditions than the B form and has the bases tilted 20° away from the perpendicular axis and slightly laterally displaced (see Fig. 1.1).

5. *The helix of double stranded DNA may be further coiled to form positive or negative supercoils.* When DNA is not supercoiled it is said to be relaxed. Positive supercoiling implies futher coiling in the direction of rotation of the existing helix, and negative supercoiling indicates coiling in the opposite direction which tends to result in an untwisting of the molecule. Enzymes known as topoisomerases are necessary for the formation or removal of supercoiling. Class II topoisomerses (also termed gyrases), together with energy from ATP hydrolysis, can convert relaxed DNA to supercoiled form. They do so, not by physically turning the molecular screw, as it were, but by nicking and resealing overlying strands so that torsion is introduced (see Fig. 1.2). Topoisomerase class I enzymes resolve

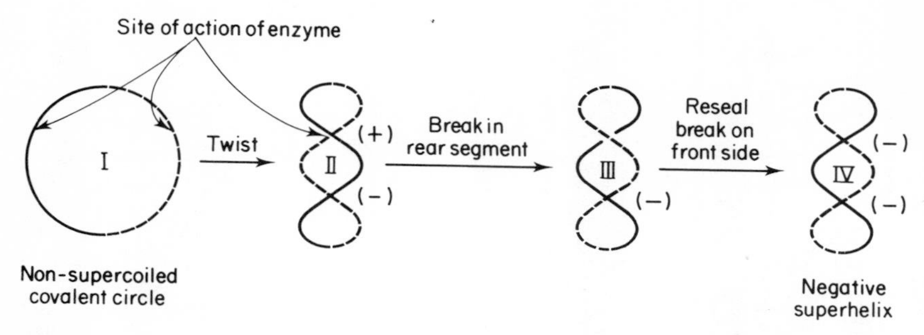

Fig. 1.2 Action of the topoisomerase II molecule of *E. coli* (also termed DNA gyrase) in introducing negative superhelix formation in a covalent circle of DNA. (From D. Freifelder, *Molecular Biology*, Van Nostrand Reinhold with permission).

supercoiling by nicking and resealing, thus establishing a return to relaxed DNA, but no energy input is required for this reaction.

6. *Some bases in DNA are modified by methylation after initial DNA synthesis.* Methylation is the only major modification undergone by DNA, and it occurs in most, but not all, organisms. In bacteria it involves chiefly the modification of adenine to 6-methyladenine, and in eukaryotes modification of cytosine to 6-methyl cytosine. It is enzyme mediated, but irreversible for a particular DNA molecule.

7. *In the DNA duplex only one strand functions as a template for messenger RNA.* The template strand is termed the sense or anticoding strand (since the mRNA message functions as the code for the protein sequence), while the sister strand with which it is base paired is the non-sense strand. However, it does not follow that the same strand of the DNA duplex is the sense strand along the length of an entire chromosome. On the contrary, it is quite clear that both strands of DNA in an entire chromosome function as sense strands in certain loci. This implies that some are read in one direction and some in another, since the two strands lie in antiparallel array and a gene can only be read in its 3′ to 5′ orientation, giving message synthesised from 5′ to 3′. So if we imagine that a chromosome's worth of DNA is spread out before us a duplex running from left to right, some genes will be read from one strand running from left to right, others from the opposite strand running from right to left. But in eukaryotes no examples exist of genes actually overlapping so that opposite strands might serve as portions of the sense strand for two quite separate messages. It is the positioning of the promoter sequences that initially determines which parts of which strands are 'read' by the polymerase.

8. *DNA molecules can be readily dissociated into single strands and then recombined by specific base pairing under appropriate conditions.* The separation of a DNA duplex into single strands is known as denaturation or melting, and its reconstitution from single stands to duplex form as renaturation or

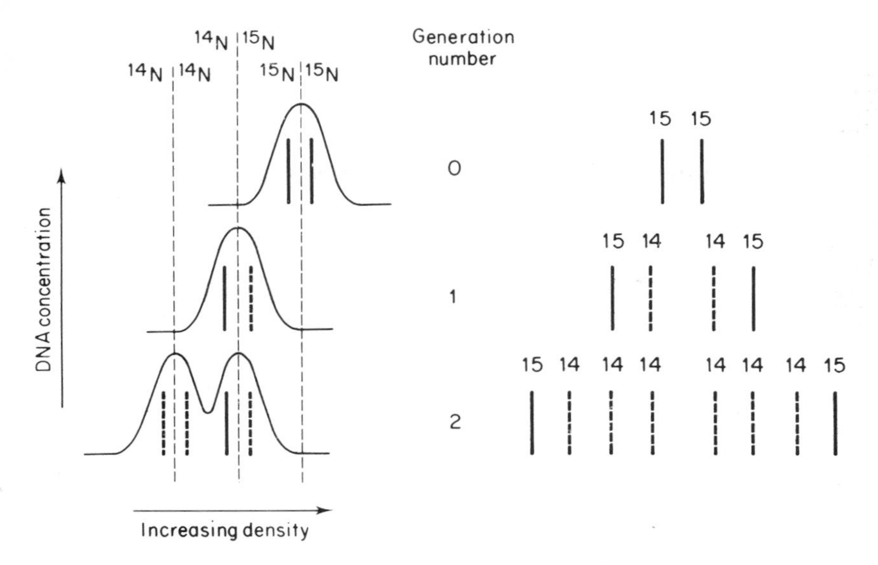

Fig. 1.3 The Meselson and Stahl experiment in which *E. coli*, grown for many generations in ^{15}N medium (black), were transferred at zero time to ^{14}N medium (hatched). After one generation, all DNA had adopted an intermediate density indicative of a hybrid molecule. In the second generation, two fractions were detected, one of ^{14}N duplex, and one of hybrid duplexes. Such an experiment confirms the semi-conservative mode of DNA replication. (From D. Freifelder, *Molecular Biology*, Van Nostrand Reinhold with permission).

annealling. If two strands from distinct sources are annealled, it is termed DNA hybridisation. The ability of the double helix to open out at least partially into a single stranded conformation is probably necessary for transcription, just as it is for replication (DNA synthesis). As we shall see later, this property of annealling makes it easy to compare sequence homologies in different strands in test tube experiments since most of the methodology of gene manipulation relies on DNA hybridisation.

9. *DNA replication is semi-conservative*; i.e. both single strands of the parent molecule serve as a template for the synthesis of new strands, and each new duplex molecule consists of one original and one newly synthesised strand. This was clearly demonstrated in the classic experiment of Meselson and Stahl (1958) outlined in Fig. 1.3. As in transcription, so in replication, the template strand is read from 3′ to 5′ and the newly synthesised strands therefore run from 5′ to 3′. As seen in Fig. 1.4, the opening up of the duplex during replication is gradual, which renders it necessary for one strand to be copied discontinuously (the so-called lagging strand), while the other can be read continuously (the leading strand). DNA replication is further complicated by the fact that a short RNA primer is synthesised first during each synthetic initiation step, to be cut out by enzymatic means later

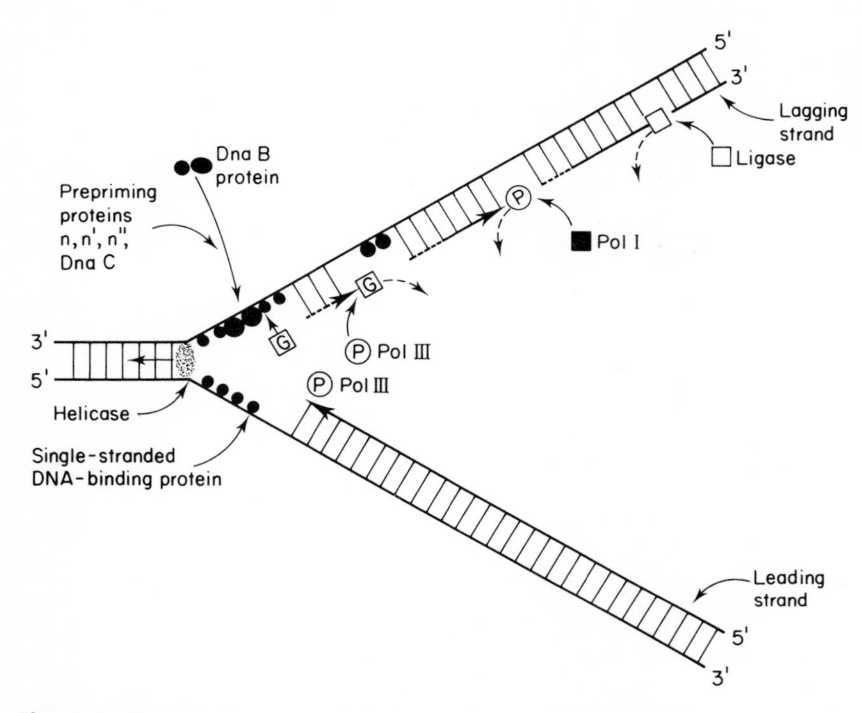

Fig. 1.4 Diagram illustrating the probable events at a bacterial DNA replication fork. Each section of DNA synthesised is headed by a short sequence of RNA primer which is removed prior to ligation of the fragments. Synthesis can only proceed in a 5′ to 3′ direction, reading the template strand from 3′ to 5′. This permits continuous replication to proceed with one strand (the leading strand) but requires discontinuous replication of the other strand (the lagging strand). (From D. Freifelder, *Molecular Biology*, Van Nostrand Reinhold with permission).

and replaced with DNA. Also the discontinuously synthesised strand produces short sequences called Okazaki fragments, and these have to be enzymatically ligated together to form a complete strand.

10. *Although most DNA exists as a linear duplex, short palindromic stretches may adopt short alternative three-dimensional structures.* As seen in Fig. 1.5, a palindromic sequence, or inverted repeat, has a central axis of symmetry, and permits base paired loops to project out from the main duplex. Although palindromic sequences are relatively abundant in DNA, it is not clear how frequently these alternative structures do actually form. They might well provide easy recognition sites for other molecules, such as regulatory proteins, which interact specifically with DNA.

11. *Many genes contain introns.* One of the most astonishing discoveries of recent years, in the field of molecular genetics, is that most gene sequences

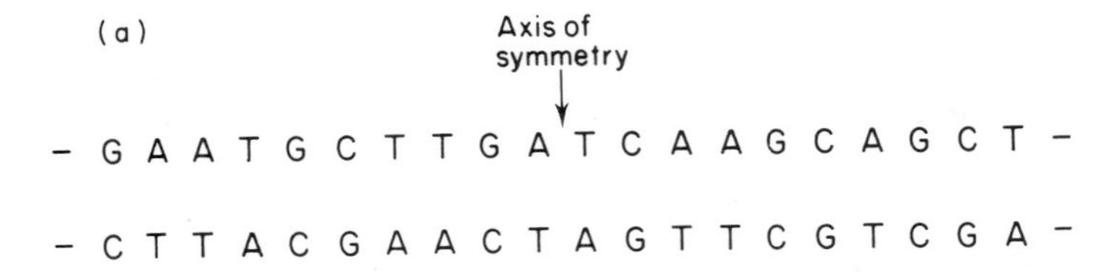

Partially palindromic sequence in normal double helical formation.

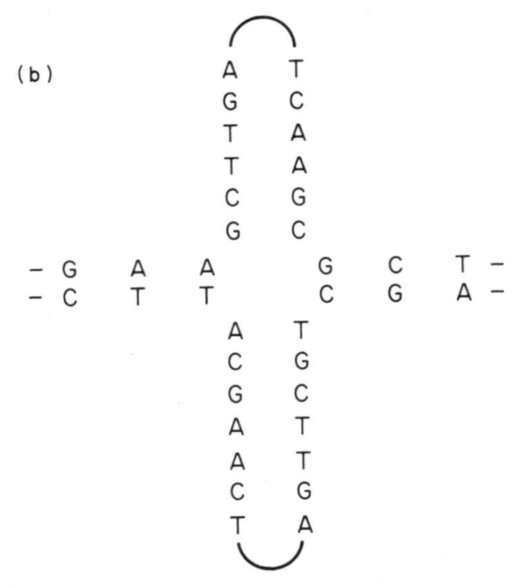

Partially palindromic sequence showing adoption of alternative conformation with specific three dimensional configuration.

Fig. 1.5 Adoption of a novel tertiary structure by a DNA sequence which includes a short palindrome.

of eukaryotic organisms (fungi, plants and animals) do not consist of a single continuous coding sequence which is simply transcribed into messenger RNA. Instead, it has become clear that most eukaryotic genes contain between one and fifty *introns*; these are lengths of sequence not represented in the message. The remaining parts of the gene which are represented in the message are known as *exons*. It is thus evident that cells must have a way of excluding the intron coding information from the message. This is done by initially transcribing the whole sequence, introns plus exons, into a large precursor RNA molecule (the heterogeneous RNA

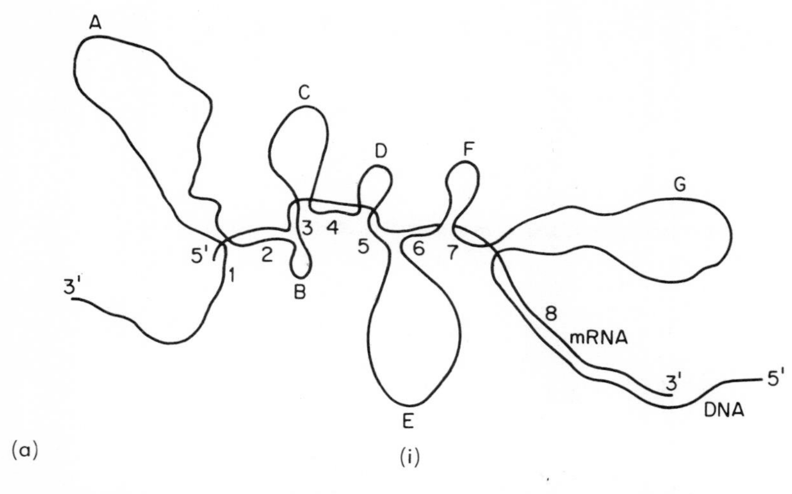

Coding sequences (exons)	Intervening sequences (introns)	Location in gene (nucleotide nos.)	Sequence size (no. nucleotides)
1		1- 47	47
	A	48-1636	1589
2		1637-1821	185
	B	1822-2072	251
3		2073-2123	51
	C	2124-2704	581
4		2705-2833	129
	D	2834-3233	400
5		3234-3351	118
	E	3352-4309	958
6		4310-4452	143
	F	4453-4783	331
7		4784-4939	156
	G	4940-6521	1582
8		6522-7564	1043
			7564

(ii)

Fig. 1.6a Reproduced with kind permission from Avers (Molecular Cell Biology) 1986. Tracing drawn from an electron micrograph of a hybrid complex between DNA (light line) and mRNA (heavy line). The DNA is a single copy of the chicken ovalbumin gene which consists of a 7564 nucleotide sequence inclusive of 7 introns. The eight exons of the gene are paired up with complementary sequences in the mRNA, while the 7 introns, not represented in the message, occur as unpaired loops in the DNA.

Fig. 1.6b The chicken conalbumin gene, which consists of 17 exons (shaded) and 16 introns. The initial transcript, hnRNA, is precisely complementary to the gene and contains an identical order of exons and introns. The mRNA, however, contains only the exon sequences of the hnRNA, the intron sequences having been spliced out in the nuclear processing of the transcript.

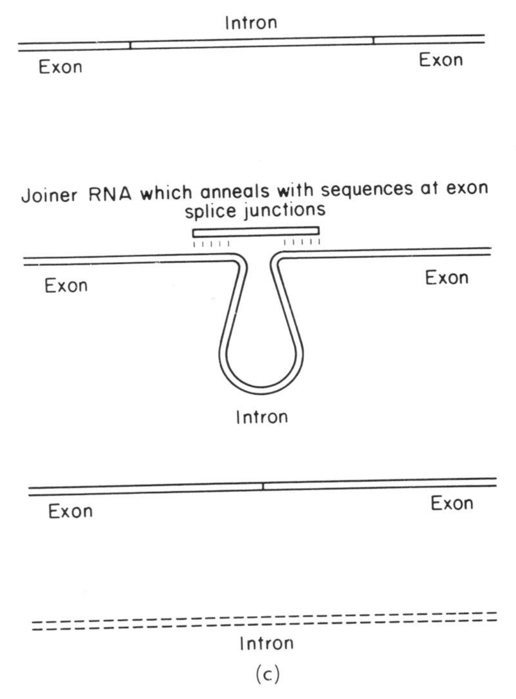

Fig. 1.6c Diagram showing a possible mechanism for intron excision from hnRNA.

or hnRNA found so abundantly in cell nuclei). This hnRNA is then modified by very precise intron excision and subsequent splicing together of the remaining exon sequences, to yield the mRNA which characterises the cytoplasm (see Fig. 1.6).

It is still not absolutely clear why higher cells are lumbered with such a complex processing arrangement to produce message. Perhaps introns are useful in protein evolution by permitting novel combinations of exons to be rapidly constructed. Interestingly, no introns have been found in bacterial cells although it may be that they were originally present early in evolution and have now been lost.

The eleven aspects of DNA structure outlined above are fundamental to understanding how DNA functions genetically and I would urge all readers of this short book to take time to master them fully. Armed with this portion of molecular biology, one will not easily lose one's way in the complex patterns of gene arrangement and the no less intricate mechanisms of gene regulation.

Two other aspects of the biological role of DNA will be summarised before a return is made to considering genes in their biological context. Firstly, a consideration of the evidence that DNA is indeed the genetic material, and secondly an account of the characteristics of the genetic code inherent in the structure of DNA. Both these aspects are considered in turn below.

1.3 Evidence that DNA is the genetic material

1. *The DNA content of sperm and egg cells in the same organism and species is constant.* Since tissues other than the cells of the germ line are frequently involved in division, the DNA content of different cells in the same organism can vary by up to a factor of two (and polyploid cells by much more). But sperm and egg cells are haploid, having only a single copy of the genome, and their DNA content is very constant, suggesting some important biological role in inheritance for the DNA.
2. *Chromosomes appear to be involved in inheritance, and consist of an approximately 50/50 mixture of DNA and protein.* Thus the composition of chromosomes does not in itself pinpoint DNA as the genetic material but does tends to disqualify RNA.
3. *The optimal mutagenic effect of ultraviolet light is at 260 nm wavelength, a value which coincides precisely with the maximal absorption of light in the absorption spectrum of the nucleic acids* (See Fig. 1.7). This suggests that mutations, which are permanent genetic changes, may involve nucleic acids, although on this basis either RNA or DNA could be the genetic material.
4. *The experiments of Avery Macleod and McArty in 1944,* followed the earlier observations of Griffiths in 1928, and established that bacteria can be permanently transformed, that is they can acquire new genetic qualities by exposure to DNA from genetically distinct bacterial strains. This work did not absolutely exclude protein from the race, since the DNA samples purified in these experiments for the induction of transformation were contaminated with protein: however, the experiments did make it seem likely that the 'transforming principle' was indeed DNA.
5. *The work of Hershey and Chase in 1952 provided the crucial proof.* These workers were able to show by radiolabelling that, when bacteria were infected with bacteriophage, only the DNA of the phage need enter the cell,

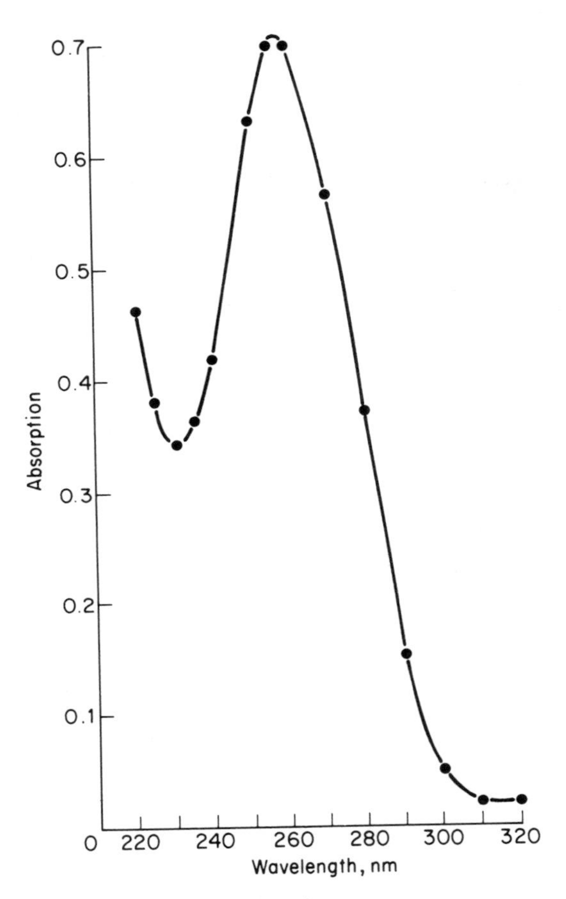

Fig. 1.7 Absorption spectrum of *Bacillus megatherium* DNA, showing low absorption at 230 nm, high at 260 nm, and a sharp fall off thereafter. Most proteins absorb maximally at 280 nm. The precise wavelength of maximal absorption of DNA varies slightly with its base composition.

yet intact phage could be recovered later from the infected cells. So in this situation, DNA alone was sufficient. In recent work, eukaryotic organisms, such as mice and fruit flies, have been genetically transformed following injection of purified cloned DNA sequences into the fertilised egg.

6. *Recognition of the double helix by Crick, Watson and Wilkins in 1953* and of the arrangement of bases in the molecule, helps to explain the functioning of the genetic code in DNA. Later work with synthetic polynucleotides by Nirenberg and Matthei in 1961 established that particular triplet codons can specify an individual amino acid in the test tube, indicating that the DNA code is indeed translated into protein structure. Therefore the gulf between genotype (the DNA) and phenotype (the protein) had been finally bridged.

1.4 The characteristics of the genetic code in DNA

1. *It is a triplet code*, in that a sequence of three bases in the DNA codes for one amino acid in the polypeptide chain of the protein molecule.
2. *The code is non-overlapping.* This statement simply means that no base in the code is shared between two or more coding triplets, so that each triplet reads on sequentially from the last. Thus, as seen in Fig. 1.8, a mutation involving a single base will affect only a single amino acid, and such has indeed been proved to be the case for many single base changes.

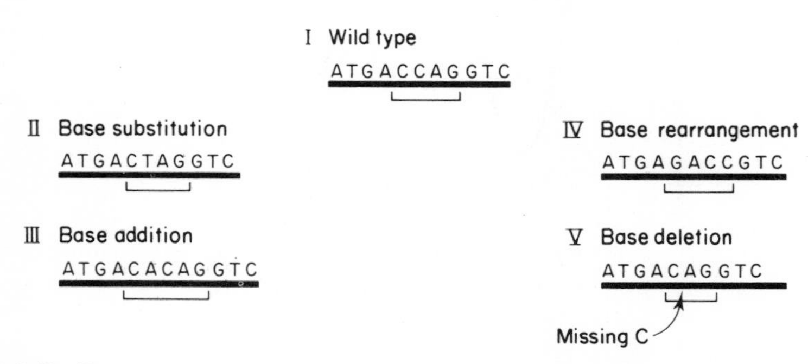

Fig. 1.8 Diagram showing four types of mutation within a confined region of a 'wild type' DNA sequence. (From D. Freifelder, *Molecular Biology*, Van Nostrand Reinhold with permission).

TT.GAG.CTA.GGA.GTA.GCT.ATT.GAC.CAG.TTT.TTA.GTA.GTT.TTT.GAG.GAT.GTA.AAG.GGC.ACT.G
GA.GGG.GAG.TCC.AGA.GAT.TTG.CCT.TCA.GGT.CAG.GGA.ATT.AAT.AAC.CTG.GAC.AAT.TTA.AGG.G
AT.TAT.TTG.GAT.GGC.AGT.GTT.AAG.GTA.AAC.TTA.GAA.AAG.AAA.CAC.CTA.AAT.AAA.AGA.ACT.C
AA.ATA.TTT.CCC.CCT.GGA.ATA.GTC.ACC.ATG.AAT.GAG.TAC.AGT.GTG.CCT.AAA.ACA.CTG.CAG.G
CC.AGA.TTT.GTA.AAA.CAA.ATA.GAT.TTT.AGG.CCC.AAA.GAT.TAT.TTA.AAG.CAT.TGC.CTG.GAA.C
GC.AGT.GAG.TTT.TTG.TTA.GAA.AAG.AGA.ATA.ATT.CAA.AGT.GGC.ATT.GCT.TTG.CTT.CTT.ATG.T
TA.ATT.TGG.TAC.AGA.CCT.GTG.GCT.GAG.TTT.GCT.CAA.AGT.ATT.CAG.AGC.AGA.ATT.GTG.GAG.T
GG.AAA.GAG.AGA.TTG.GAC.AAA.GAG.TTT.AGT.TTG.TCA.GTG.TAT.CAA.AAA.ATG.AAG.TTT.AAT.G
TG.GCT.ATG.GGA.ATT.GGA.GTT.TTA.GAT.TGG.CTA.AGA.AAC.AGT.GAT.GAT.GAT.GAT.GAA.GAC.A
GC.CAG.GAA.AAT.GCT.GAT.AAA.AAT.GAA.GAT.GGT.GGG.GAG.AAG.AAC.ATG.GAA.GAC.TCA.GGG.C
AT.GAA.ACA.GGC.ATT.GAT.TCA.CAG.TCC.CAA.GGC.TCA.TTT.CAG.GC.C.CCT.CAG.TCC.TCA.CAG.T

Fig. 1.9 Nucleotide sequence of part of the 'early' region of the genome of SV40 virus. The reading frame for translation, reading from the top left corner is -TT; GAG; CTA; and so on, and the sequence codes for the T antigen protein of the virus.

3. *The code is comma-less, but not unpunctuated.* There is no information in the code indicating the beginning or end of a coding triplet (the anti-codon) so that if one or two bases are added, the whole reading frame may shift and all subsequent amino acids will be altered. However, an addition or deletion of

three bases leads to no change in reading frame. The punctuation present in the code consists of specific triplets which act as stop signals, serving to denote the end of an individual polypeptide chain in the messenger RNA molecule. It is therefore inaccurate to describe the genetic code as lacking in punctuation. But, as seen in the example shown in Fig. 1.9, with the exception of stop signals, there are no visible breaks in the code word sequence in the DNA. Promoter sequences and initiation points, which will be described in subsequent chapters, serve to instruct the RNA polymerase molecules as to where the coding sequence of a gene actually starts.

4. *It is a degenerate code.* Degeneracy refers to the fact that some amino acids may be specified by more than one anti-codon triplet of bases. This follows from the fact that a code of four different letters can generate a series of 64 different triplet anti-codons, yet only twenty amino acids need be specified. Three triplets are used to indicate termination but the other 61 are freely available. The variability in the anti-codon triplet in the transfer RNA molecules determining one amino acid resides almost entirely in the last base, and there seems to be some flexibility of fit of the tRNA molecule anti-codon and the codon triplet in the messenger RNA. Such flexibility is assumed in the 'wobble hypothesis' of Crick. So there are two distinct kinds of degeneracy relevant to the use of the genetic code. In the first, a single tRNA molecule can base pair with more than one codon, and in the second, more than one species of tRNA exists for certain amino acids. Both kinds of degeneracy occur.
5. *It is an interrupted colinear code.* The colinearity of the genetic code refers to its relationship with the amino acid sequence in the protein specified. The order of codons in the message (and anti-codon triplets in the DNA) is in the same linear order as are the amino acids specified by these codons in the polypeptide chain. But the colinearity is dramatically interrupted by the existence of intron sequences in the DNA, so that only the exon sequences are truly colinear with the amino acid sequences; thus some base triplets that are distant in the DNA (because they straddle an intron) determine amino acids that are adjacent and directly colinear in the polypeptide chain.
6. *The code is quasi-universal.* Its universality stretches from bacteria to the human, so that bacterial genes can be effectively processed and translated into the appropriate protein product if introduced into the nuclei of human cells, or human genes into bacterial cells. So the same anti-codon triplets specify the same amino acids in both organisms. The only exceptions are to be found in the cellular organelles, the mitochondria, in which various evolutionary changes have occurred in the particular amino acids determined by certain triplets. Thus the term quasi-universal.

Following this brief consideration of DNA structure and function, it is now appropriate to move on to the main topic of this book, the gene, and to see how these units are arranged in the genomes of animals and plants. Those readers interested in a fuller account of the molecular biology of DNA and its coding characteristics are referred to in *The Genetic Code and Protein Biosynthesis*, by B.F.C. Clark, Edward Arnold, 1984.

1.5 Analyzing the Genome

1.5.1 The C value paradox

As seen in Table 1.1, the haploid amounts of DNA in various organisms are highly variable, and there are a number of quite distinct aspects to this variability. Firstly, it is apparent that there is a tendency for the C value (the amount of DNA in the haploid genome, as in a sperm cell) to increase in line with the progression of evolutionary complexity. For example, an insect, such as *Drosophila*, has approximately 0.1 picograms (1 picogram = 10^{-12} g), fish have between 0.5 and 3 (with some exceptions to be discussed shortly) and the human about 3. Since 1 picogram of DNA contains about 10^9 base pairs, it means that in our own species the entire available coding material is 3×10^9 bases. If we imagine than an average gene codes for a protein of 300 amino

Table 1.1 DNA content of haploid genomes expressed in picograms per cell (1 picogram = 10^{-12} gm). 1 picogram of DNA is equivalent to approximately 1×10^9 base pairs.

Common Name	*Latin Name*	*DNA Quantity*
Phage	Bacteriophage lambda	0.00005
Phage	Bacteriophage T4	0.00017
Virus	SV40 virus	0.000005
Bacterium	Escherichia coli	0.004
Yeast	Saccharomyces cerevisiae	0.024
Fungus	Neurospora crassa	0.017
Paramecium	Parmecium aurelia (polyploid)	160.0
Alga	Chlamydomonas reinhardi	0.06
Maize	Zea mays	3.9
Onion	Allium cepa	16.8
Lily	Lilium longiflorum	36.1
Pea	Pisum sativum	4.5
Silkmoth	Bombyx mori	0.52
Fruitfly	Drosophila melanogaster	0.18
Crab	Cancer borealis	1.7
Rainbow trout	Oncorhynchus mykiss	2.5
Tench	Tinca tinca	0.9
Lung fish	Protopterus aethiopicus	50.0
Xenopus	Xenopus laevis	3.15
Frog	Rana pipiens	7.6
Salamander	Amphiuma means	84.0
Alligator	Alligator mississipiensis	2.5
Chicken	Gallus domesticus	1.25
Mouse	Mus musculus	2.5
Human	Homo sapiens	3.0

acids, that suggests a requirement for 900 bases. But, as remarked earlier, most genes are interrupted by introns, and an average gene is some 5 times longer than its combined exon regions as a result of the presence of introns. This would suggest an overall length of 4500 bases per gene, and if regulatory regions upstream and some spacer region downstream are added, we might conclude that a gene's worth of DNA might be, on average, about 5000 bases. If we divide 5000 bases into the human coding potential of 3×10^9 bases, it appears that some 600 000 genes could be accomodated. However, it can be calculated from data regarding mutation rates, as well as extrapolations about protein diversity, that probably not more than 50 000 genes are active in the human genome. So less than one tenth of the genome appears to be actually required as gene coding material. This is the first aspect of what has come to be called the C value paradox, namely that as evolutionary complexity demands a larger genome, the increase in genome size more than outstrips this requirement for extra coding material. As we shall see later in this chapter, the types of sequences which occupy much of the apparently redundant DNA are at least partially known, and include regulatory sequences, pseudogenes, and a proportion of perhaps functionless DNA.

However, a second puzzle is in store, if Table 1.1 is scrutinised further, namely that some organisms have extravagantly more DNA than others, even as compared to closely related species. Examples include *Protopterus*, the African lung fish, *Amphiuma* a North American Salamander, and *Lilium longiflorum*, as eastern member of the plant genus *Lilium*, all of which have more than ten times the total amount of DNA present in other species of fish, amphibian, or dicotyledonous plant to which they must be closely related. So the C value paradox refers also to this apparent anomaly. The explanation for this astonishing variation is still far from clear, especially since the species with exceedingly large genome sizes do not seem to have gained more of some special fraction of DNA or particular type of sequence; rather they just seem to have more of all types of sequences. Yet they are not simply polyploid, and so the total number of active gene sequences in *Amphiuma* is presumably no greater than in *Rana* (the frog).

1.5.2 Not all DNA is made up of the same kinds of sequences

One way of trying to understand the curious features of the C value paradox is to study the characteristics of DNA more closely, in order to deduce whether it really is all the same material. Biochemically there is no doubt that it is, but if a way could be devised to study the sorts of sequences made up by the bases, something meaningful might emerge. Happily, a rather simple aspect of DNA biochemistry has helped to provide a handle to this problem. It is that DNA duplexes can be reduced to a single stranded form by heating (since the hyrogen bonds between sister strands are broken), and that, when the temperature is returned to a more moderate level, single strands will anneal with one another; not necessarily with the precise strand with which they were originally complexed but with any other strand of a complementary base sequence. Such melting and annealling experiments can be used to provide quite detailed information about what sorts of sequences occur in a particular genome, and how abundant any one sequence or type of sequence actually is.

What is done in practice is to purify a sample of DNA from a convenient tissue derived from the chosen organism. This sample is then sheared by sonication, or by some other appropriate means, in order to reduce the molecular mass and so convert the small number of enormously long DNA molecules into a larger number of a more manageable size. Size is important since the molecules must interact in the hybridization reaction, and must be able to move around readily in the solution in order to facilitate finding a complementary strand. Once the DNA solution has been converted to one containing solely molecules of a few hundred bases long, it can be placed in a conveniently heated receptacle for the experimental thermal denaturation or melting. This receptacle is usually a quartz vial in an ultraviolet spectrophotometer, since in this instrument the melting and annealling procedures can be followed by absorption of UV light at 260 nm wavelength. When DNA duplexes come apart by breakage of the hydrogen bonds between the bases (as occurs at elevated temperatures) an increase in UV absorption occurs. This is the so called hyperchromic effect, since the single strands absorb more (an increase of approximately one third) 260 nm light than do the equivalent duplexes. When the DNA reassociates in suitable conditions, the absorption once more declines, and so this elevation and reduction in absorption provides a neat and easy way of monitoring the behaviour of the DNA.

The sample of DNA, now reduced to relatively small fragments of a few hundred bases each, and suspended in a salt solution containing approximately 0.1 M sodium phosphate is gently heated from the 25°C ambient to almost 100°C. A cap on the vial ensures that water is not lost in the heating process by evaporation. For most samples, a dramatic increase in absorption will have occurred between 80°C and 100°C, raising the absorption from 1 unit to approximately 1.3. The temperature at which a sample of DNA melts from the duplex to the single stranded form is a result of various factors, of which the most important is the base composition of the DNA. Since the GC base pair is held by three hydrogen bonds and the AT pair by only two, DNA samples which are especially rich in GC will have a somewhat higher melting temperature. In a eukaryotic genomic sample the strand separation temperature will vary with variant base composition and a range of melting temperatures will be obtained. Following prolonged exposure to 100°C, to ensure complete denaturation, the sample is rapidly cooled and then held at a steady temperature of about 65°C. At this temperature, given an optimal ionic environment, the DNA strands will reassociate by the complementary base pairing, and the 260 nm absorption will decline appropriately. But this reassociation is for the most part quite slow and the absorption may continue to decline for some hours. It is also interesting to note that, for reasons that will be made clear shortly, a return to the baseline absorption will not be obtained with DNA samples from eukaryotic sources.

Fig. 1.10 shows a representation of the sort of reassociation curve that would be expected for DNA from a complex eukaryote. What can we learn from this

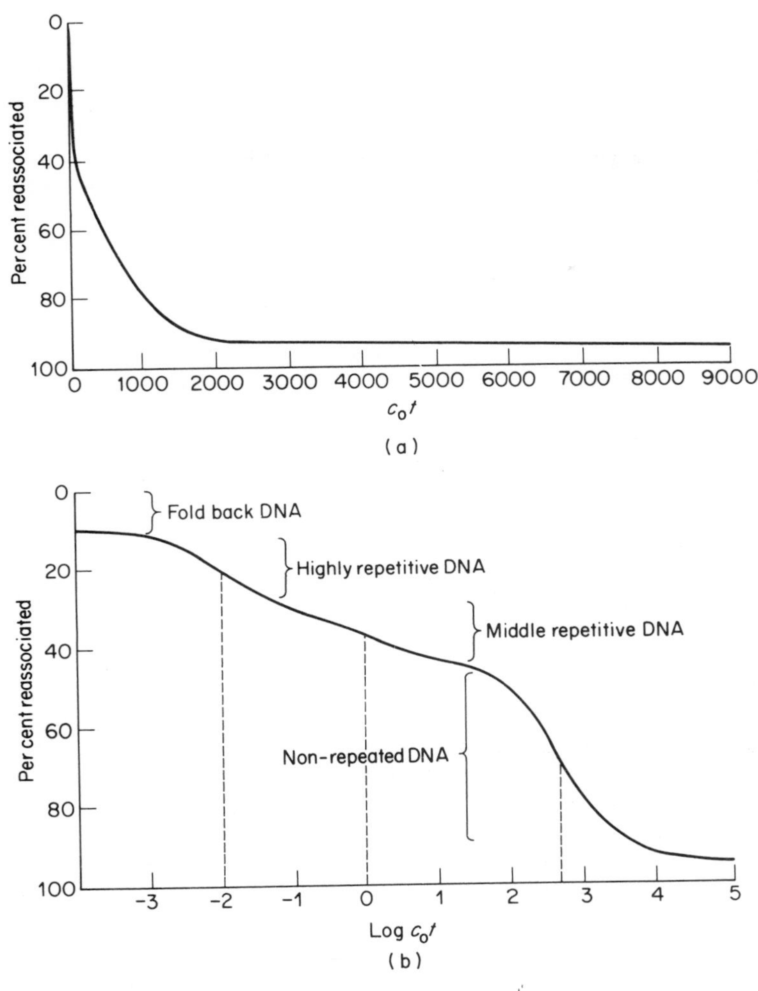

Fig. 1.10 The profile of molecular reassociation of human DNA. DNA has been purified, reduced in molecular weight by mechanical shearing, and following its thermal dissociation, is held at 65°C in 0.24M sodium phosphate. The graph in (a) represents the amount of reassociation (measured by reduction in optical density at 260 nm) expressed on the vertical axis, as a function of DNA concentration × time (indicated as C_0t), and expressed on the horizontal axis. The contributions to the reaction made by the four main sequence classes in the DNA is most easily seen where the log of C_0t is taken as the horizontal axis, and this is shown in (b). (Adapted from Schmid, C.W. and Deininger, P.L. (1975) *Cell*, **6**, 347).

curve, and in particular, what does it tell us about the types of DNA sequences present in the sample?

Firstly we should consider the constraints in the DNA solution on DNA annealling. Imagine a DNA solution which has been melted out from only one species of DNA sequence, say a viral circular genome. On melting, only two kinds of molecule are present in solution and so each strand has a 50/50 chance of finding a complementary strand first time around. Alternatively, if a DNA solution consists of a few million molecules, each of which is unique, there is only one chance in millions that a single strand will find its complement first time. So it will be clear that the time taken for a molecular sequence to reassociate can be taken to be a measure of its frequency in the solution. Indeed, as it turns out, single copy gene sequences scarcely reassociate at all in normal DNA renaturation experiments.

A second important parameter in these experiments is the concentration of the DNA. It will be clear on reflection that if a DNA sample is very dilute, then even a sequence with numerous sister strands in the same sample may take a long time to associate due to the diffusion times involved. Thus, in general, the more concentrated the sample, the faster the reassociation times. In practice concentration and time are both taken together in moles per litre and seconds to give the horizontal axis of the reassociation curve, the vertical axis being percentage reassociation as determined by 260 nm absorption. Such a curve is known as a *Cot curve*.

The reassociation curve can be seen to consist of four separate curves which combine to yield the overall melting profile. Starting from the time of first incubation at 60 °C, a small fraction of the DNA sample anneals instantly. This material is known as *foldback DNA* and consists of sequences which are inverted repeats, also known as palindromic sequences. As seen in Fig. 1.11, such a sequence is able to snap back on itself to form a hairpin like structure, actually a duplex with two free ends. The second component occupies a greater range of Cot values, and results from the relatively rapid reassociation of DNA known as *highly repetitious DNA*. Such material is made up of sequences of between 10 and 1000 nucleotides that are present, clustered in each genome as at least 10^4 copies of each sequence. So it is readily understandable that each of these sequences will hybridise quickly at a low Cot value.

A third component of the curve, occupying an even greater range of Cot values than the last fraction, reflects reassociation of DNA sequences that are *moderately repetitious* – although the precise boundary between this fraction and the highly repetitious fraction is somewhat arbitrary – it is made up for the most part of sequences that belong to families. The members of one of these families are similar, but not necessarily identical, and hybridise readily with other sequences in the same family, but not those outside the family. They are generally present in 100s or low 1000s of copies per haploid genome. The last component accounts for the remainder of the material, and is made up of sequences present once or as very few copies per genome. This fraction is known as *unique sequence* or *single copy* DNA, although neither title is an entirely accurate description of the material present. We should note that a proportion of genomic DNA from eukaryotic sources will not hybridise in normal Cot curve experiments, no matter how long the experiment is continued, and so some strands persist as single strands and the original 260 nm absorption figure is never quite regained.

1.5.3 The distribution of different types of sequence in the genome

The question posed at the beginning of Section 1.5.2 above can therefore be answered in the negative. Genomic DNA is not all the same, and quite distinct types of sequences can be detected by a Cot curve analysis. Where in the genome are these different fractions found, and what functions do they fulfil?

Before determining where the different fractions occur in the genome, it is useful to know how much of each type of sequence is present in a single genetic complement. Foldback DNA accounts for only 1–2% on average, highly repetitious sequences about 10%, moderately repetitious anything from 20–30% and the 'single-copy' DNA approximately 50%: however, this can be substantially higher in some lower eukaryotes where the third fraction is considerably smaller. In prokaryotes virtually all of the DNA is unique sequence.

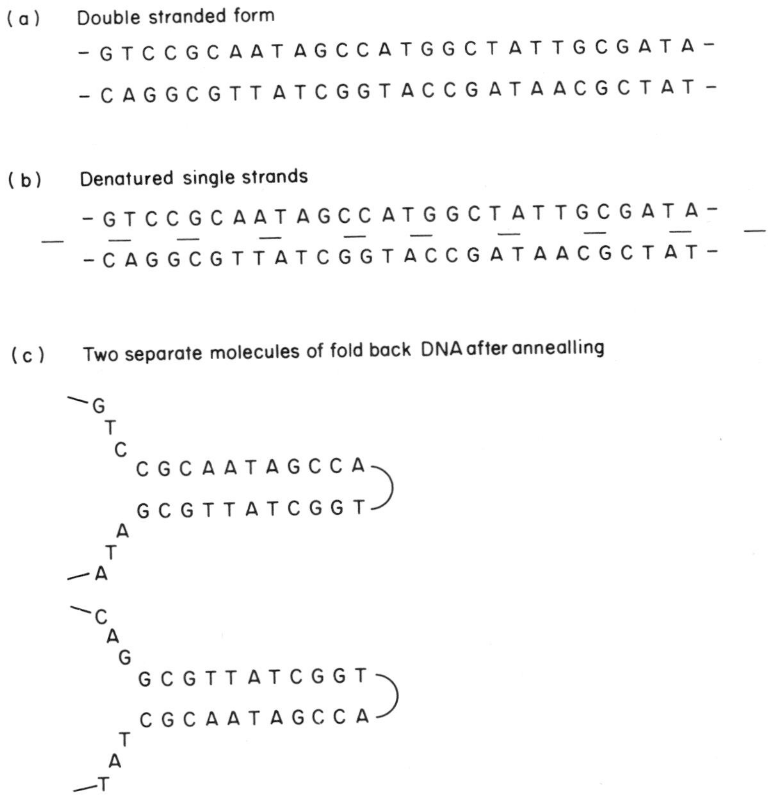

(a) Double stranded form

- G T C C G C A A T A G C C A T G G C T A T T G C G A T A -
- C A G G C G T T A T C G G T A C C G A T A A C G C T A T -

(b) Denatured single strands

- G T C C G C A A T A G C C A T G G C T A T T G C G A T A -
- C A G G C G T T A T C G G T A C C G A T A A C G C T A T -

(c) Two separate molecules of fold back DNA after annealling

G T C C G C A A T A G C C A
A T G G C G T T A T C G G T

C A G G C G T T A T C G G T
T A T C G C A A T A G C C A

Fig. 1.11 The 'fold-back' phenomenon which occurs when palindromic sequences within duplex molecules are converted by denaturation to single strands. Such fold-back renaturation is extremely rapid when suitable renaturation conditions are provided.

Foldback DNA

This rather peculiar type of sequence is present most abundantly in regulatory regions lying upstream from the gene coding regions. They permit DNA to adopt novel three-dimensional structures and these may allow recognition of these sites by DNA polymerase molecules or other enzymes which interact with the DNA.

Highly repetitious sequence DNA

This material consists for the most part of tandemly arranged repeat sequences, each some hundred bases long, although in some species, such as the crab *Cancer*, a considerable fraction of DNA consists of a string of only two repeat bases, A and T. Since about 10% of the genome consists of this material, the sequences within this fraction are present as at least 10^4 copies per genome, sometimes much more. The base composition of this material is often quite different from the bulk of the DNA of the cell, permitting its isolation as a DNA satellite when genomic DNA, after sonication to break it up into fragments, is centrifuged to equilibrium on a caesium chloride density gradient. Not all highly repetitious DNA is recoverable as a satellite, only certain sequences in certain species.

Although the function of this highly repetitious sequence DNA remains obscure, much of it is concentrated around the chromosomal centromeres, and it may play some structural role in chromosome pairing. Blocks of highly repetitious sequence DNA occuring in sites other than the centromeric regions are said to be intercalary, as we shall discuss in Chapter 3. Since this type of DNA is quite distinct and exists in very condensed chromatin, it gives rise to a chromatin subfraction known as constitutive heterochromatin. It is rarely transcribed into RNA.

Moderately repetitious sequence DNA

This material, varying as it does in amount within different organisms, also varies greatly within itself. Repeat sequences present between ten and a few thousand times are likely to fall into this category as a result of their renaturation kinetics, with many of these sequences belonging to distinct families of similar, but non identical sequences. Members of the same family will hybridise, but members of differing families will not. Included here, then, are sequences which are gene coding regions for transfer RNA, ribosomal RNA, and proteins such as histone. Many other moderately repetitious sequences are not gene coding regions but may be regulatory; it is possible that some of the promoter regions of genes coding for message will fall into this category although in many cases the homologous regions will be short. This is largely due to the fact that all such genes are transcribed by one type of polymerase enzyme, called RNA polymerase II, and all the promoters must be recognised by one protein and can therefore be expected to contain some consensus sequence regions. As discussed in Chapter 4, the CAT and TATA boxes are such consensus sequences.

However, not all the moderately repetitious material can be accounted for within the types of sequence mentioned, and most of it appears not to be transcribed and to have no obvious function (although some repetitious sequences without obvious function are, however, found to be transcribed and are therefore represented in the hn RNA fraction). Indeed in the human, one family of sequences, the *Alu* family (named after the restriction endonuclease enzyme that cleaves it into two) comprises about 3% of the total genome and consists of between 300 000 and 500 000 copies of sequences which are only about 300 base pairs long. This material is classed as moderately repetitious because the repeats are not tandemly arranged but are interspersed between other sequences, and thus do not renature as speedily or as completely. At least some of the *Alu* sequence DNA is transcribed and it has been proposed that it fulfils a special role in replication of DNA.

It will be clear, then, that this third type of DNA is very variable and comprises a great assortment of different types of sequence, some well defined, others still of entirely unknown function.

Unique sequence DNA

It must be emphasised once more that the term unique is not intended to be taken literally in all cases, since some of the sequences occur as families of a few sequences. But it comprises for the most part sequences understood to be the true genes, that is sequences of DNA which uniquely carry the code for individual protein molecules. The number of different proteins in a higher eukaryote is probably between 20 000 and 50 000, although some of these are determined by slightly repetitious sequences. Thus proteins such as globin, the protein moiety of the respiratory pigment haemoglobin, crystallin, the protein which makes up the bulk of the cell content of the fibre cells of the lens of the eye, and actin, the major component of muscle and contractile mechanisms in cells, are all examples of gene products existing in variety. Table 1.2, shows at least six

Table 1.2 Human haemoglobins and their subunits. Each haemoglobin molecule is a tetramer of four globins.

Haemoglobins	*Symbols*	*Globin subunits*
Embryonic haemoglobins	HbE_1	$\alpha_2\epsilon_2$ (Gower 2)
	HbE_2	$\xi_2\epsilon_2$ (Gower 1)
	HbE_3	$\xi_2\gamma_2$ (Portland)
Foetal haemoglobin	HbF[a]	$\alpha_2\gamma_2$
Adult haemoglobins	HbA	$\alpha_2\beta_2$
	HbA_2	$\alpha_2\delta_2$

[a]The human γ chain gene is duplicated and the two copies are not absolutely identical. There are, therefore, two types of human HbF, differing in the amino acid at the 136 position.

different human globins, each the product of a different gene sequence. In addition, there may be a number of identical or near identical copies of one sequence coding for one globin, as indeed there are for the human gamma globin. The globin gene family has another fascinating feature in store for us, namely the existence of pseudogenes. When gene cloning technology permitted a search of genomic DNA for sequences that hybridised to a globin probe, some sequences appeared which were very similar to globin genes but were not represented in the protein repertoire of the organism. Pseudogenes are often defective in a number of ways, having introns missing (then being termed a processed pseudogene and assumed to result from reverse transcription of messenger RNA) and, most frequently, having non-open reading frames by dint of possessing many stop codons mixed up within the run of a normal globin sequence. Such genes are assumed to be an accidental inclusion in the genome, but nevertheless, appear as part of a 'unique sequence' gene family in hybridization. Processed pseudogenes occur randomly throughout the genome (as expected in the reverse transcription hypotheses) whereas others are usually close to functional genes.

The reason for so many of the protein coding genes occurring as small families of sequences is presumably due to the duplication of one original ancestral sequence in the course of evolution, a theme I will return to in Chapter 6.

1.5.4 Sequence interspersion in the genome

The position of certain sequences or groups of sequences in the genome has already been indicated, namely the concentration of highly repetitious short sequences in the vicinity of centromeres. Some other sequences are clustered, as are for example the sequences coding for large ribosomal RNA in the region of the nucleolus. But one of the most remarkable aspects of the partitioning of distinct DNA sequences within the genome is the marked tendency for unique sequences to be interspersed between repetitious sequences. The precise manner of this interspersion pattern varies in different eukaryotes, but in most cases, repeated sequences of between 300 and 600 base pairs alternate with non repeated sequences of between 800 to 2000 base pairs. At least half the entire genome of many organisms consists of DNA arranged in such an interspersed pattern, although in some groups, for example insects, the interspersion pattern is different, chiefly in that the tracts of repetitious DNA between the unique sequences are much longer.

It is therefore clear that the enormous length of genomic DNA, carrying as it does only four different bases, can none the less be subdivided into a considerable range of different kinds of material, differing only in the arrangement of the bases, but having very different evolutionary origins and fulfilling quite different functions within the cell.

2

Small Genomes

In this Chapter the characteristics and regulation of genomes of viroids, viruses, plasmids, mitochondria, chloroplasts and bacteria will be considered. Since few bacteria have genome sizes in excess of 0.05 picograms (1 picogram = 10^{-12} gm which is roughly equivalent to 1×10^9 base pairs of DNA) the range being considered is from 5×10^7 base pairs for larger bacteria down to as little as 5386 bases for phage *phi* X174, a single stranded DNA phage. As set out in Table 2.1, there are marked differences between the genomes of prokaryotes and eukaryotes, such as the absence of histones, nucleosomes and chromatin from the former. But although most of the particles here being considered have no true chromatin, some viruses which reside in the nuclei of eukaryotic cells, such as SV40, have DNA complexed with histone in the form of nucleosomes (see Fig. 2.1). Such histones are cellular in origin, yet the resulting mini-chromosomes of these viruses not only exist in the cell but are packaged in this form into the protein capsid of the virion.

Of the genomes to be considered, most are circular but some are linear, quite a few are single rather than double stranded nucleic acid, and some are formed from RNA rather than DNA. Although viroids, the first to be briefly mentioned, seem to be as simple as it is possible to be in terms of a molecular replicating unit, there are other factors, known as prions, the causative agent of the mysterious disease of sheep, scrapie, which seem to consist 'only particles' of protein and no nucleic acid can be recovered from the infective agent.

2.1 Viroids

These factors are pathogens of higher plants, causing conditions such as potato spindle tuber disease. Although mainly of academic interest in genetic terms, they are remarkable in having a single stranded RNA genome which has closed ends and, as seen in Fig. 2.2, is substantially base paired. No protein or other molecular structure has been found to be associated with these RNA molecules.

Table 2.1 Features that distinguish prokaryotic from eukaryotic cells. (From Maclean, N. and Hall, B.K., *Cell commitment and Differentiation*, Cambridge University Press, 1987, with permission).

Feature	*Prokaryotic*	*Eukaryotic*
Size	mostly 1–10 μm	most 10–100 μm
Multicellular forms	rare	common, with extensive tissue formation
Respiration	many strict anaerobes (oxygen fatal) facultative anaerobes and aerobes	all aerobic, but some facultative anaerobes by secondary modifications
Metabolic patterns	great variation	all share cytochrome electron transport chains. Kreb's cycle oxidation, Embden-Meyerhof glucose metabolism
Flagellae	simple structure composed of the protein flagellin	complex 9 + 2 structure of tubulin and other proteins
Photosynthetic enzymes	bound to cell membranes as composite chromatophores	enzymes packaged in plastids bounded by membrane
Sexual systems	rare; if present one-way (and usually partial) transfer of DNA from donor to recipient cell occurs	both sexes involved in sexual participation and entire genomes transferred; alternation of haploid and diploid generations is also evident
Genetic material	double-stranded DNA: genes not interrupted by intron sequences (thought by some to have been lost in course of evolution)	double stranded DNA: genes frequently interrupted by intron sequences, especially in higher eukaryotes
Plasmids	commonly present	rare
Chromatin with histone	–	+
Nucleus	–	+
Nuclear membranes	–	+
Cellular organelles:		
Mitochondria	–	+
Endoplasmic reticulum	–	+
Vacuoles	–	+
Lysosomes	–	+
Chloroplasts	–	+ [a]
Centrioles	–	+ [b]
Ribosomes	+ (70S)	+ (80S)
Microtubules	–	+
Cell membrane	+	+
Cell wall	present on most but not all cells	present on plant and fungal cells only

[a]Only in plants.

[b]Absent from higher plants.

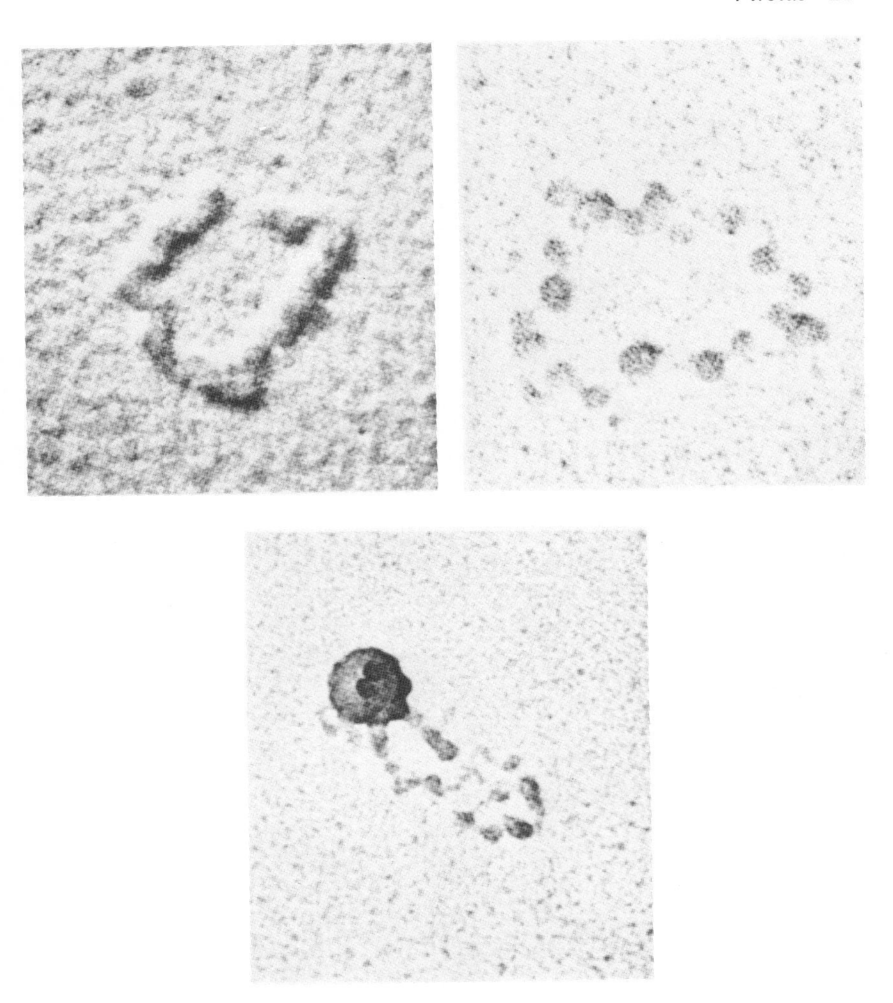

Fig. 2.1 The mini chromosomes which result from the packaging of the genome of the animal virus SV40 into a nucleosomal conformation. That at top left in an electron microscope view of a spread chromosome fixed in 0.15M NaCl and that at top right in 0.01M NaCl. The fact that the chromosome remains intact within the virion is revealed in the lower picture, which shows extrusion of the mini-chromosome from an SV40 virion × 500 000. (From Kelly, T.J. and Nathans, D., Adv. Virus Res., **21**, 86, (1977). Photographs kindly provided by Professor D. Nathans).

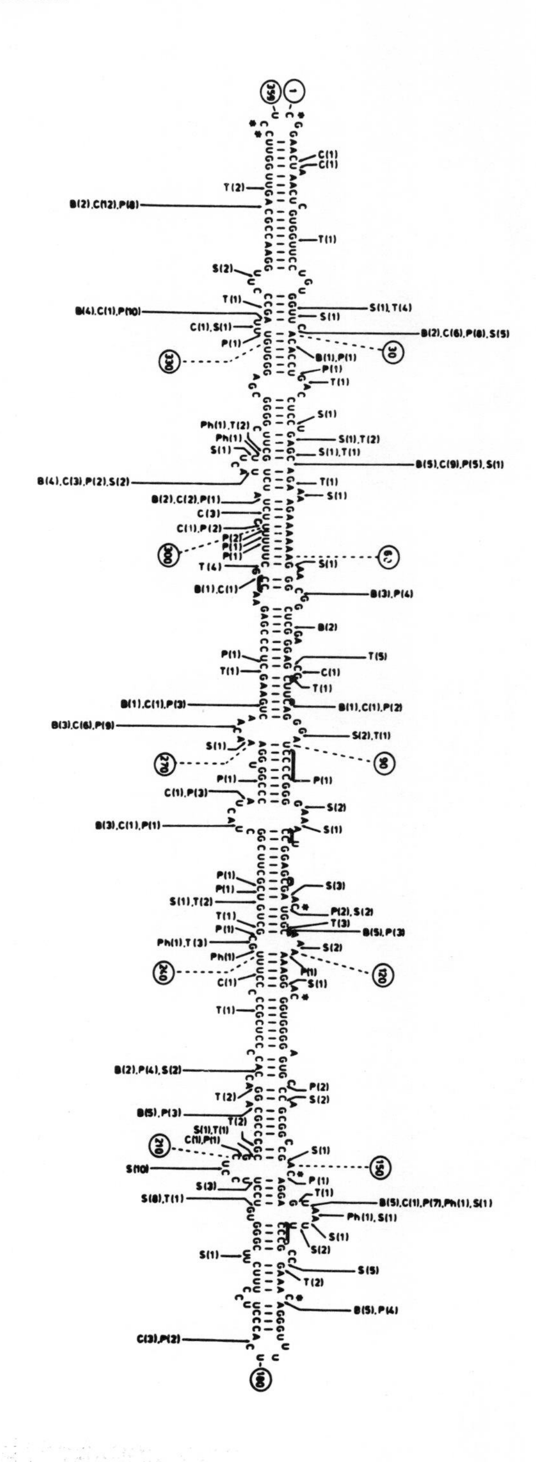

Table 2.2 Distribution of the four types of nucleic acid in the genomes of viruses. (From Dunimock, N.J. and Primrose, S.B., *Introduction to Modern Virology*, 3rd edition, Blackwells, 1987 with permission).

	Distribution		
Type of nucleic acid	*Bacterial viruses*	*Plant viruses*	*Animal viruses*
Single-stranded DNA	Not very common. Best examples ϕX174 and filamentous phages	Rare. Only found in geminiviruses	Common
Double-stranded DNA	Most common	Rare. Only found in caulimoviruses	Common
Single-stranded RNA	Not very common. Best examples are the male-specific coliphages	Most common type found.	Common
Double-stranded RNA	Rare. Only found in cystoviruses	Only found in Reoviridae	Common

2.2 Viruses

These cellular parasitic agents come in a very great variety of forms and in this short book we can only afford space to consider a few in detail. Distinct groups of viruses are to be found infecting the cells of plants, animals and bacteria, and indeed a range of different viral types are to be found infecting all of these different organisms. So any observations about viruses based purely on the plant, animal, and bacteriophage subdivision are extreme generalisations. However, Table 2.2 indicates that within these broad viral categories, some trends can be detected on the basis of the type of viral genome present. We can see that four distinct genomic categories exist, namely those having single stranded DNA, single stranded RNA, double stranded DNA or double stranded RNA. All four genomic types are found commonly in different groupings of animal viruses, while most plant viruses have single stranded RNA genomes and most bacterial viruses have double stranded DNA genomes. But examples displaying each genomic type can be found in each grouping of organisms. It should also be borne in mind that some viruses multiply in both plant and invertebrate hosts.

Since viruses are so diverse, the genetic features of these factors are also exceedingly varied, and detailed discussion of all types and all the peculiar

Fig. 2.2 The probable structure of the potato spindle tuber viroid. It is a single stranded closed circle of DNA, but substantially base paired throughout its length. Loops are presumed to form where base pairing is precluded by non complementary bases in the opposite strands. From Gross, H.J., *et al. Nature*, **273**, 203 (1978). Figure kindly supplied by Dr. H.J. Gross.

genetic features associated with them is quite beyond the scope of this slim volume. What is presented is a relatively detailed account of a few selected examples, followed by a discussion of some of the more remarkable genetic features found in some of the others. In this way it is hoped that reasonable coverage will be achieved.

2.2.1 Simian Virus 40 (SV40)

This is a small skewed icosahedral virus of vertebrate cells. It is a member of the papovaviruses and has been entensively studied, its genome being known in considerable detail. The closed circular genome is of double stranded DNA, a length of approximately 5 kilobase pairs, and the DNA is complexed with nucleosomes, as shown in Fig. 2.1, so the genome is actually in the form of a minichromosome within the virion and in the nucleus of an infected cell. Both strands of the DNA are used as sense strands, but there is a marked difference in the timing of transcription of the two strands. One strand is transcribed early in the replicative cycle of the virus (but only partially, since transcription does not overlap between strands), the proteins synthesised from this early RNA being necessary for viral DNA replication and also for the later transcription. At least three proteins result from the late transcription, but the mRNA for one of these is a partial transcript of the mRNA for another. A third aspect of interest in SV40 gene regulation is the use of leader sequences, that is a 200 base length section of 16s RNA molecule coding for a protein product derives from a DNA sequence which is non adjacent in the genome, thus indicating that the leader RNA is spliced to the rest of the message post-transcriptionally.

A final point of interest concerning the SV40 genome is the presence of *enhancer sequences*, series of bases in the DNA which although not part of a gene sequence, can profoundly affect the efficiency of transcription of an individual gene sequence in the cell nucleus. Such enhancers are located adjacent to and either before or after the sequence which they affect. Enhancer sequences have also been discovered in eukaryotic genomes and are discussed again in Chapter 5. So the SV40 genome is characterised by four distinct and interesting aspects

- a minichromosome,
- subdivision into early and late transcription,
- leader sequences,
- enhancer sequences.

2.2.2 Retroviruses

This class of animal viruses contains viruses with icosahedral nucleocapsids with an envelope of plasma membrane origin. The genome consists of two separate but similar long strands of single-stranded RNA, associated with two very short strands (see Fig. 2.3) the latter molecules actually being cellular

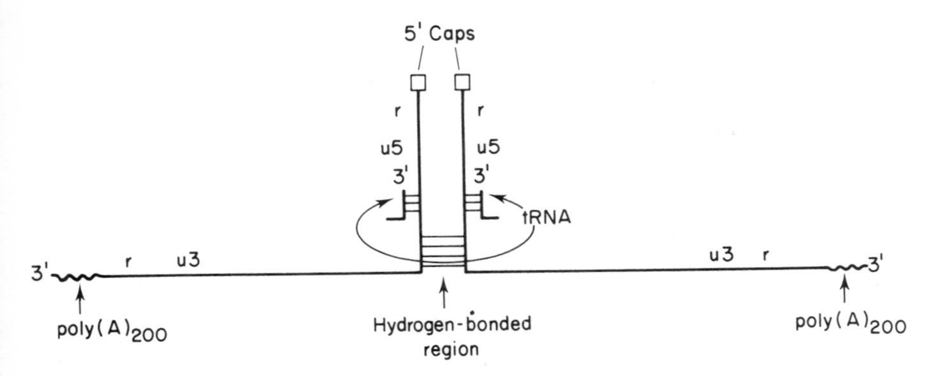

Fig. 2.3 Structure of a retrovirus genome within the virion, as exemplified by Rous sarcoma virus RNA. Two long RNA molecules are hydrogen bonded over a short region but such that the strands lie in parallel rather than antiparallel array. Thus the 5′ ends are at the same ends of the two molecules. Poly A tails are present on both 3′ ends of these long RNAs, and two molecules of tRNA are also partially base paired to the structure. These latter molecules act as primers for replication.

transfer RNA molecules which act as primers for replication. Retroviruses have a number of important genetic attributes. Firstly, they are exceedingly simple, each RNA carrying only four genes, coding for a protein called gag which forms a viral core protein (the gag gene), a reverse transcriptase enzyme (the pol gene), an envelope protein (the env gene) and a gene which codes for an oncogenic transforming protein (the onc gene). At each end of this four gene sequence is a long terminal repeat sequence of about 560 bases.

Retroviruses, by their possession of a pol gene which codes for a reverse transcriptase, are able to make a DNA copy of their own RNA template. This unique property, ultimately leading to production of a double stranded DNA copy, allows the retrovirus to integrate into the DNA genome of the invaded cell as a provirus. Such integration is not like the integration of phage lambda into the bacterial DNA, which is an option which lambda may or may not choose. Instead, integration of the retroviral provirus into the host DNA is essential for its own transcription and further replication.

One of the genes previously denoted on the retroviral genome was a so called oncogene. Not all retroviruses carry oncogenes and not all are oncogenic. For example, the AIDS and other lenti viruses (these include a number of viruses which infect the brain of sheep and other mammals), are non-oncogenic retroviruses. But many retroviruses are oncogenic, owing this property to their possession of oncogenes. A range of about twenty different oncogenes are known to be carried by retroviruses, each strikingly similar to the sequence of a normal eukaryotic gene (the so called protooncogene) from which they are presumably derived. The *modus operandi* of oncogenes in cancer induction is not understood, but is believed to result from disturbance of the regulation of the protooncogenes, many of which code for kinase enzymes or growth factors.

2.2.3 Tobacco mosaic virus (TMV)

This well studied plant virus is placed in a grouping called the Tobamoviruses. It consists of a straight tubular filament in which the protein subunits are arranged helically to form the hollow tube, and the single stranded RNA genome is wound around the inside of the cylinder. This virus featured in a classic experiment carried out in 1957 by Fraenkel-Conrat and Singer. It had been found that TMV virus with full infectivity could be reconstituted by permitting the self-assembly of genomic RNA and disaggregated protein subunits from the viral coat. These workers selected two distinguishable strains of this virus, and in both cases mixed the genomic RNA of one with the protein capsomere subunits of the other. What became evident was that when the reconstituted virus was used to infect plants, the form of the virus propagated was, in each case, determined solely by the RNA and not the protein.

The RNA genome of TMV is translated into a series of three proteins, two of which are made from partial overlap, since one of the start codons occurs within the coding sequence of the first protein. The third protein, known as coat protein which forms the outer capsomeres ofthe virion's coat, is translated from a terminal portion of RNA and does not seem to be effectively translated from the virion genomic RNA of the virus, but from newly synthesised RNA in the plant cell. The arrangement of these genes is shown diagrammatically in Figure 2.4.

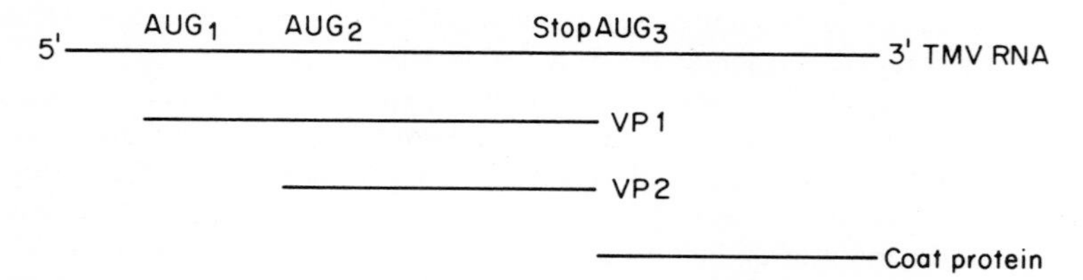

Fig. 2.4 The RNA genome of tobacco mosaic virus showing how it is differentially translated into three separate proteins. Three separate AUG start codons are present, with a single stop codon preceding the third start. From D. Freifelder, *Molecular Biology*, Van Nostrand Reinhold with permission.

2.2.4 Bacteriophage T4

As seen in Figure 2.5 T4 is a classical large phage with an elongated icosahedral head, a contractile tail, and a series of kinked tail fibres. It is an *E. coli* phage and has an optimal infective cycle from cell infection to cell lysis of only 30 minutes. The T4 genome is a linear molecule of double stranded DNA, although as will

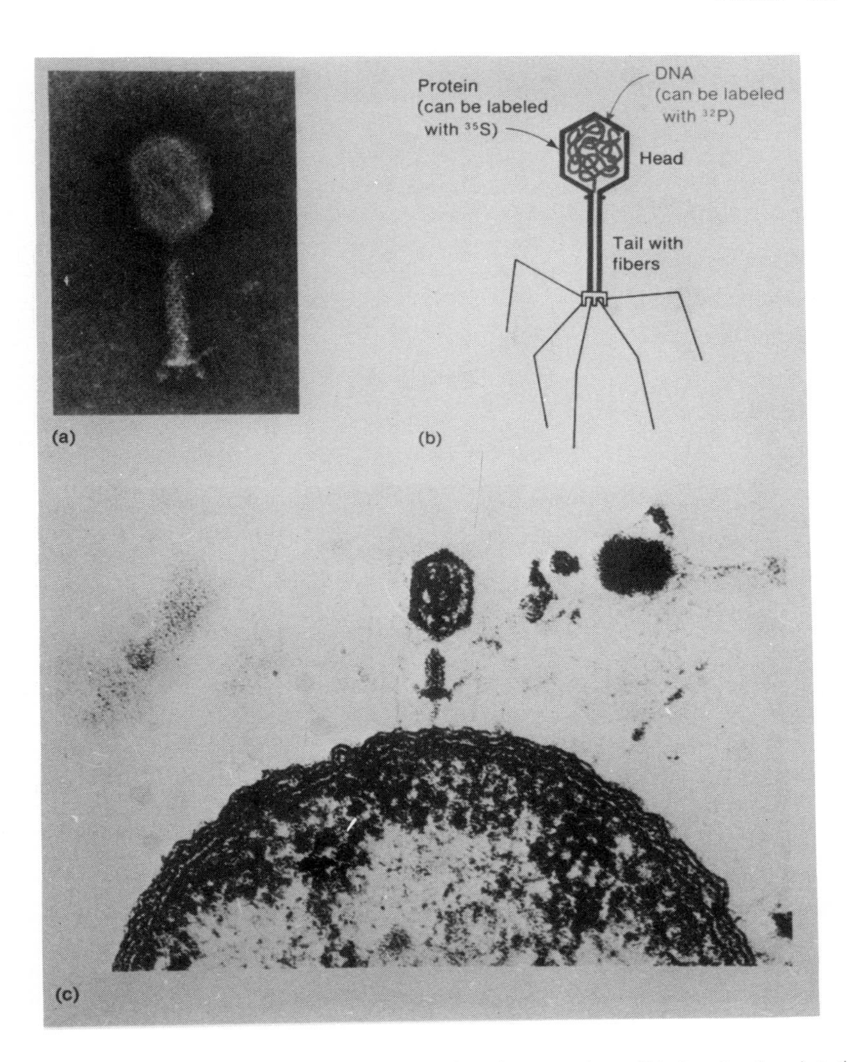

Fig. 2.5 (a) Electron microscope picture of the bacteriophage T4 showing head, tail, and base plate × 150 000. (b) Diagram of the basic structure of T4. (c) Electron micrograph showing a T4 phage adsorbed to the surface of an *E. coli* cell. The micrograph shows a section through the bacterial cell × 120 000. From D. Freifelder, *Molecular Biology*, Van Nostrand Reinhold with permission. (a) Courtesy of E. Kellenberger and (c) Lee Simon and T. Anderson.

be discussed in a moment, it may also exist in a circular form. The DNA has a total linear length of about 60 μm, or 166 000 base pairs, and some 50 different genes have been identified. A partial map is shown in Fig. 2.6, and, as this diagram indicates, the genetic map is best presented in circular form.

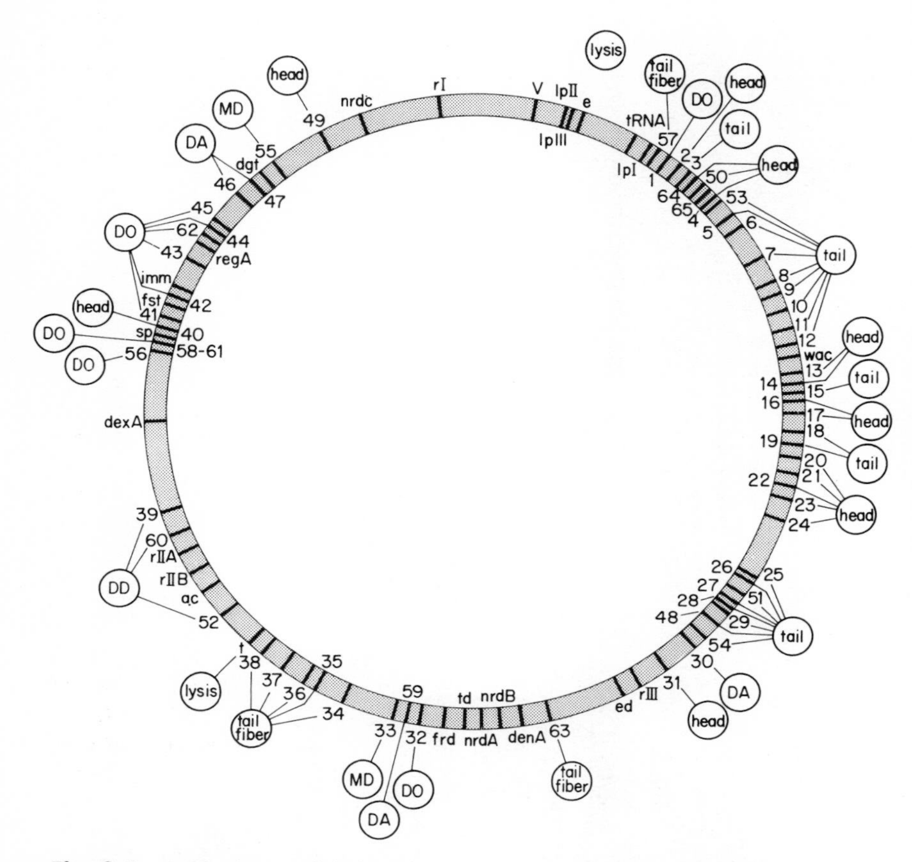

Fig. 2.6 A genetic map of the bacteriophage T4. Some gene loci are numbered, having been scored by mutational analysis, others are named where products have been identified. Mutants which are defective for certain activities are located as DO (no DNA synthesis); DA (arrested DNA synthesis); DD (delayed DNA synthesis); and MD (maturation defective). (From Lewin B. *Gene Expression*, **3**, Wiley (1977) with permission).

Phage T4 has a number of interesting and unusual genetic features which will now be considered. Firstly, in common with other phages such as T2 and T6, its DNA contains the unusual base hydroxymethylcytosine as a replacement for cytosine. (Some other phages replace thymine with uracil or hydroxymethyluracil). This curious feature of the T4 genomic DNA has a rather fortunate advantage, namely that its replication cycle within a bacterial cell can be rather easily followed by the appearance of this base in the DNA, since bacterial DNA does not contain hydroxymethylcytosine.

Secondly, if the gene order for different phages is determined, the sequence of genes is found to be *circularly permuted*. This explains why it has a linear genome but a circular genetic map. So different individual phages have different genes at the ends of the molecules but all have the same linear order. The molecules are also *terminally redundant*, that is the same genes are repeated at each end of the linear duplex. This is explained by the fact that the genome is longer than is necessary to include only one copy of each gene, and, since the order is the same in each case, the sequences at the ends are repeated. This also implies that the ends can be made 'sticky' by exonuclease digestion, which removes single stranded DNA from either end; such sticky ended molecules can actually be circularised in the test tube.

The final aspect of T4 genetic organisation which merits comment is the fact that the temporal expression of the genome can be divided, as can that of the much simpler genome of SV40 virus, into early and late transcription, – relative to the replicative cycle, they are now subdivided into pre-early, early and late. A further facet of this temporal segregation is the clustering of genes controlling similar functions, so that they occupy adjacent or near-adjacent locations in the genome. Thus the genes which regulate DNA replication are 'early' genes and three of these are clustered. As we might expect, early genes tend to code for products required in the later stages of the replicative cycle, while late genes code for proteins such as tail fibres.

2.2.5 Bacteriophage lambda (λ)

This phage,like T4, consists of an icosahedral head with attached tail, but no contractile elements exist in the tail and only a single tail fibre is present (Fig. 2.7). Lambda is the best known of the temperate phages, that is bacterial viruses which may indulge in autonomous replication and consequent cell lysis, or alternatively, may integrate into the *E.coli* genome as a stable prophage. It is a phage with a double stranded DNA genome of about 46 500 base pairs, enough for a least 35 genes. Of this gene complement, it is interesting to note that 20 are concerned with producing head and tail proteins, 9 with DNA replication and lysis, and 6 are regulatory.

Lambda DNA can exist in linear or circular forms, but differs from T4 in that both forms occur naturally during vegetative replication. The linear form has naturally cohesive (sticky) ends and so will readily circularise. Indeed only lambda DNA with cohesive ends is infective.

In one of its cellular forms the lambda genome may exist as twisted and partially supercoiled molecules which are converted to open circles by nicking. The interconversion of the different forms of lambda DNA is illustrated in Fig. 2.8.

Phage lambda's main claim to fame is as an example of a virus with alternative lytic and lysogenic pathways, and it is this aspect which will now be considered. On entering a bacterial cell, the question is whether the invading phage lambda will enter a lytic or lysogenic pathway? In fact, within an infected culture of *E.coli*, some do one and some do the other. It is known to involve competition between two regulatory proteins, termed lambda repressor protein

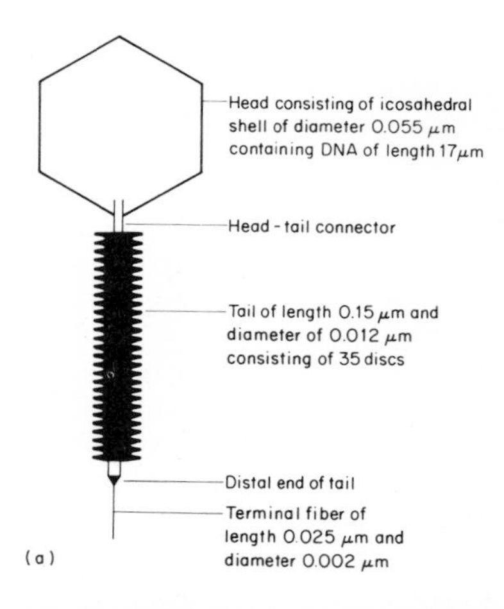

(b)

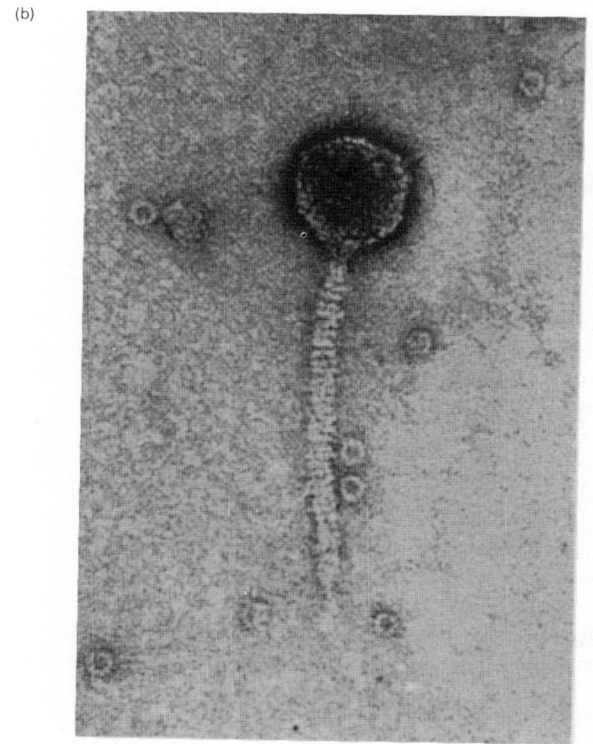

Fig. 2.7 (a) Diagram of the structure of bacteriophage lambda virion. (b) Electron micrograph of phage lambda showing head and tail. Magnification × 250 000. From Lewin B., *Gene Expression* **3**, Wiley (1977) with permission. Photograph in (b) kindly supplied by Dr. A. Howatson.

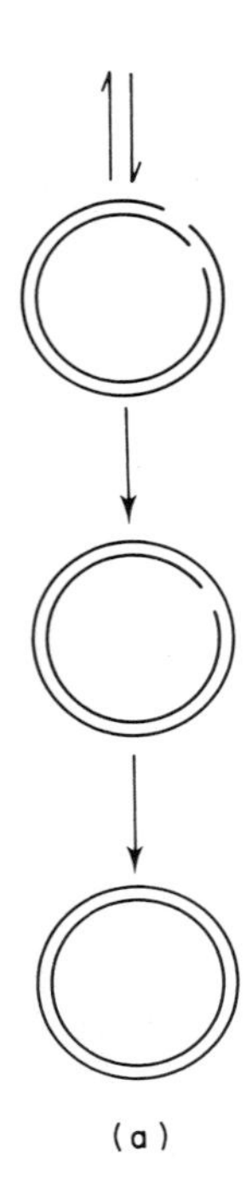

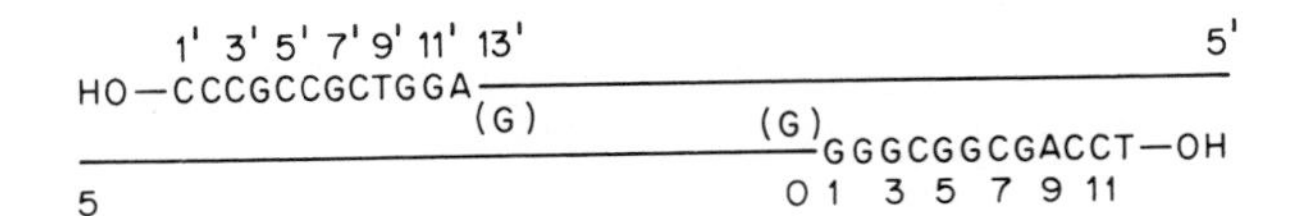

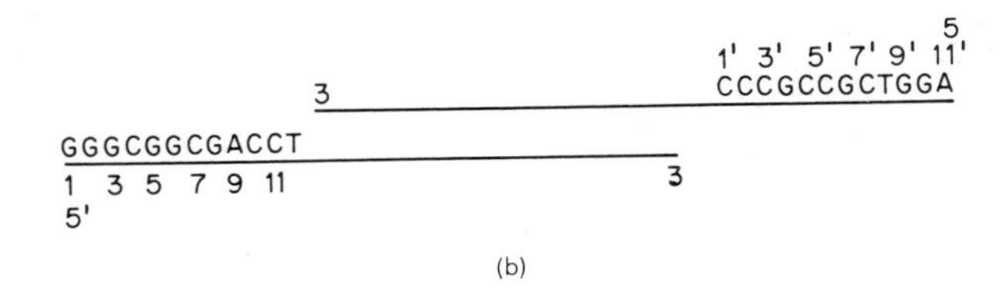

Fig. 2.8 (a) Linear and circular forms of phage lambda DNA, showing steps in the interconversion of one to the other. The protruding single stranded ends of the linear form can anneal with each other to form a partial circle, and the breaks are then sealed by enzymic action forming covalent bonds. (b) Detail of the cohesive ends of phage lambda DNA. In the lower part, the protruding cohesive ends are shown. If DNA polymerase I is allowed to 'fill in' the missing bases, as shown in the upper part, both circularization and infectivity are abolished. From Lewin B., *Gene Expression*, **3**, Wiley (1977) with permission. Data of Wu and Taylor (1971).

and CRO. The repressor protein prevents activity of genes involved in the lytic cycle, while CRO antagonizes lysogeny and permits lysis to proceed. The balance struck between the repressor and CRO protein is competition for DNA binding sites – that is further complicated by the fact that each tends to repress the synthesis of the other.

During the lysogenic cycle, the lambda genome is integrated as a prophage into the *E. coli* genome. This involves its colinear insertion by site-specific recombination. Lambda must be in a circular form to be integrated but is of course linear following insertion of the prophage DNA. We should note in passing that some other phages other than lambda are not site-specific and some

```
CTCCCGACTGCCTATGATGTTTATCCTTTGAATGGTCGCCATGATGGTGGTTATTATACC
GTCAAGGACTGTGTGACTATTGACGTCCTTCCCCGTACGCCGGGCAATAACGTTTATGTT
GGTTTCATGGTTTGGTCTAACTTTACCGCTACTAAATGCCGCGGATTGGTTTCGCTGAAT
CAGGTTATTAAAGAGATTATTTGTCTCCAGCCACTTAAGTGAGGTGATTTATGTTTGGTG
CTATTGCTGGCGGTATTGCTTCTGCTCTTGCTGGTGGCGCCATGTCTAAATTGTTTGGAG
GCGGTCAAAAAGCCGCCTCCGGTGGCATTCAAGGTGATGTGCTTGCTACCGATAACAATA
CTGTAGGCATGGGTGATGCTGGTATTAAATCTGCCATTCAAGGCTCTAATGTTCCTAACC
CTGATGAGGCCGCCCCTAGTTTTGTTTCTGGTGCTATGGCTAAAGCTGGTAAAGGACTTC
TTGAAGGTACGTTGCAGGCTGGCACTTCTGCCGTTTCTGATAAGTTGCTTGATTTGGTTG
GACTTGGTGGCAAGTCTGCCGCTGATAAAGGAAAGGATACTCGTGATTATCTTGCTGCTG
CATTTCCTGAGCTTAATGCTTGGGAGCGTGCTGGTGCTGATGCTTCCTCTGCTGGTATGG
TTGACGCCGGATTTGAGAATCAAAAAGAGCTTACTAAAATGCAACTGGACAATCAGAAAG
AGATTGCCGAGATGCAAAATGAGACTCAAAAAGAGATTGCTGGCATTCAGTCGGCGACTT
CACGCCAGAATACGAAAGACCAGGTATATGCACAAAATGAGATGCTTGCTTATCAACAGA
AGGAGTCTACTGCTCGCGTTGCGTCTATTATGGAAAACACCAATCTTTCCAAGCAACAGC
AGGTTTCCGAGATTATGCGCCAAATGCTTACTCAAGCTCAAACGGCTGGTCAGTATTTTA
CCAATGACCAAATCAAAGAAATGACTCGCAAGGTTAGTGCTGAGGTTGACTTAGTTCATC
AGCAAACGCAGAATCAGCGGTATGGCTCTTCTCATATTGGCGCTACTGCAAAGGATATTT
CTAATGTCGTCACTGATGCTGCTTCTGGTGTGGTTGATATTTTTCATGGTATTGATAAAG
CTGTTGCCGATACTTGGAACAATTTCTGGAAAGACGGTAAAGCTGATGGTATTGGCTCTA
ATTTGTCTAGGAAATAACCGTCAGGATTGACACCCTCCCAATTGTATGTTTTCATGCCTC
CAAATCTTGGAGGCTTTTTTATGGTTCGTTCTTATTACCCTTCTGAATGTCACGCTGATT
ATTTTGACTTTGAGCGTATCGAGGCTCTTAAACCTGCTATTGAGGCTTGTGGCATTTCTA
CTCTTTCTCAATCCCCAATGCTTGGCTTCCATAAGCAGATGGATAACCGCATCAAGCTCT
TGGAAGAGATTCTGTCTTTTCGTATGCAGGGCGTTGAGTTCGATAATGGTGATATGTATG
TTGACGGCCATAAGGCTGCTTCTGACGTTCGTGATGAGTTTGTATCTGTTACTGAGAAGT
TAATGGATGAATTGGCACAATGCTACAATGTGCTCCCCCAACTTGATATTAATAACACTA
TAGACCACCGCCCCGAAGGGGACGAAAAATGGTTTTTAGAGAACGAGAAGACGGTTACGC
AGTTTTGCCGCAAGCTGGCTGCTGAACGCCCTCTTAAGGATATTCGCGATGAGTATAATT
ACCCCAAAAAGAAAGGTATTAAGGATGAGTGTTCAAGATTGCTGGAGGCCTCCACTATGA
AATCGCGTAGAGGCTTTGCTATTCAGCGTTTGATGAATGCAATGCGACAGGCTCATGCTG
ATGGTTGGTTTATCGTTTTTGACACTCTCACGTTGGCTGACGACCGATTAGAGGCGTTTT
ATGATAATCCCAATGCTTTGCGTGACTATTTTCGTGATATTGGTCGTATGGTTCTTGCTG
CCGAGGGTCGCAAGGCTAATGATTCACACGCCGACTGCTATCAGTATTTTTGTGTGCCTG
AGTATGGTACAGCTAATGGCCGTCTTCATTTCCATGCGGTGCACTTTATGCGGACACTTC
CTACAGGTAGCGTTGACCCTAATTTTGGTCGTCGGGTACGCAATCGCCGCCAGTTAAATA
GCTTGCAAAATACGTGGCCTTATGGTTACAGTATGCCCATCGCAGTTCGCTACACGCAGG
ACGCTTTTTCACGTTCTGGTTGGTTGTGGCCTGTTGATGCTAAAGGTGAGCCGCTTAAAG
CTACCAGTTATATGGCTGTTGGTTTCTATGTGGCTAAATACGTTAACAAAAAGTCAGATA
TGGACCTTGCTGCTAAAGGTCTAGGAGCTAAAGAATGGAACAACTCACTAAAAACCAAGC
TGTCGCTACTTCCCAAGAAGCTGTTCAGAATCAGAATGAGCCGCAACTTCGGGATGAAAA
TGCTCACAATGACAAATCTGTCCACGGAGTGCTTAATCCAACTTACCAAGCTGGGTTACG
ACGCGACGCCGTTCAACCAGATATTGAAGCAGAACGCAAAAAGAGAGATGAGATTGAGGC
TGGGAAAAGTTACTGTAGCCGACGTTTTGGCGGCGCAACCTGTGACGACAAATCTGCTCA
AATTTATGCGCGCTTCGATAAAAATGATTGGCGTATCCAACCTGCA
```

Fig. 2.9 Half of the entire DNA sequence of the small phage ϕX174. It reads from top left to bottom right as one continuous strand, containing 5386 nucleotides and coding information for nine genes (some overlapping). (From B. Alberts, *Molecular Biology of the Cell*, Garland, (1983), with permission).

seem to be able to integrate in almost any site on the bacterial genome.

Five viruses have now been discussed and relevant aspects of their genetic mechanisms highlighted. However, other important aspects of gene organisation and function do not occur in the five viruses discussed, so before concluding discussion of viruses, a short review of certain other genetic aspects will be given.

2.2.6 Other genetic mechanisms used by viruses

Two other remarkable genetic phenomena are to be found amongst viruses, and both will be illustrated by reference to the small phage termed ϕX 174, which has a closed circle of single stranded DNA as genome. This genome codes for nine different gene products, yet has only 5375 bases. Its complete nucleotide sequence is known (Fig. 2.9). A study of Fig. 2.10 reveals that six of these genes are partially overlapping, so that the end of gene A overlaps with the start of gene C, the end of C overlaps the start of D and the end of gene D overlaps the

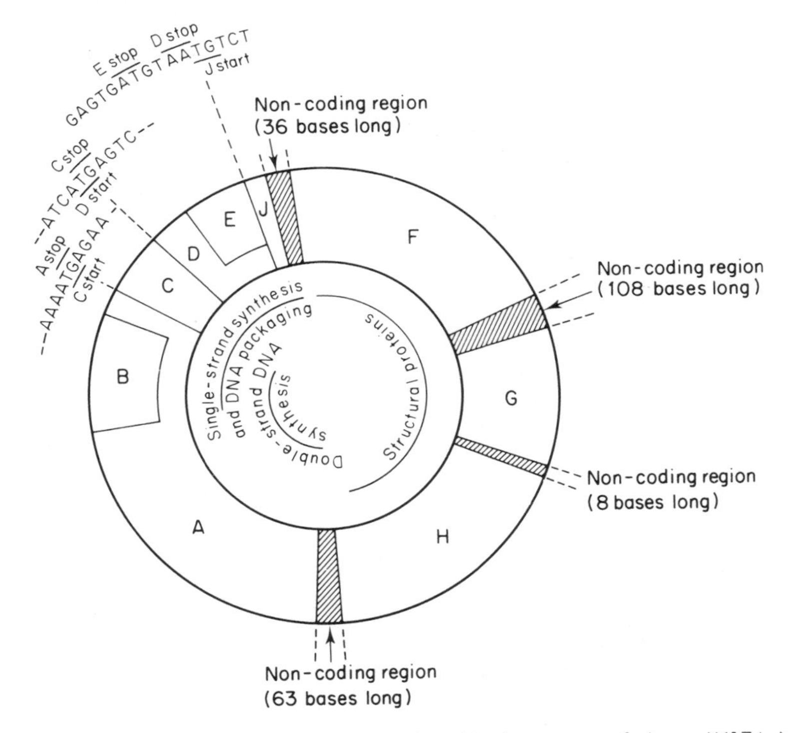

Fig. 2.10 Arrangement of the nine genes found in the genome of phage ϕX174. Non coding regions are hatched. Genes B, K and E occur entirely within areas which overlap with the coding sequences of other genes. (From Dimmock, N.J. and Primrose, S.B. Introduction to Modern Virology, 3rd. edition, Blackwells, (1987), with permission).

start of J. There are also two additional genes, B and E, that lie entirely within other genes, namely A and D respectively. How is this amazing economy of coding information regulated? The answer seems to be that some transcription is permitted to read through weak termination sequences, so ensuring the production of at least some polycistronic messages. Another is that genes within genes are read in different reading frames, thus utilising the same basic code to provide two quite distinct sets of information. Since there must be great difficulty in evolving genes within genes, in terms of the acquisition of useful polypeptide constructs, it seems that this small phage is an impressive example of what evolution can achieve given sufficient selection pressure, in this case being, at least in part, the limits placed by a packaging system on maximal use of a small genome.

In concluding this brief look at virus genomic wizardry, it is appropriate to set out in diagrammatic form the different strategies employed by different viruses to achieve a common goal – the synthesis of messenger RNA which will be read by the translational mechanisms of the parasitised cell to make more viral protein. Fig. 2.11 demonstrates the different strategies adopted by viruses with distinct types of genomes, often involving short-lived intermediate molecules, to produce mRNA, in this case styled + mRNA. This figure and our earlier discussion emphasise that viruses utilize an amazing series of genetic tricks to compensate for their tiny genomes, tricks which, by and large, are not encountered in the larger and more flexible genomes of cellular organisms.

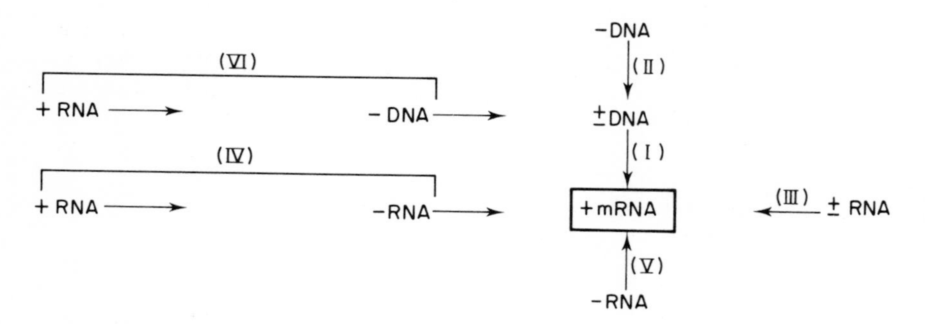

I Double-stranded DNA viruses
II Single-stranded DNA viruses
III Double-stranded RNA viruses
IV Single-stranded RNA viruses, mRNA identical in base sequence to virion RNA
V Single-stranded RNA genome complementary in sequence to mRNA
VI Single-stranded RNA genome with a DNA intermediate in their growth.

Fig. 2.11 Diagram illustrating the different pathways of messenger RNA as used by different classes of viruses, mRNA being designated as +RNA. (From Pennington, T.H. and Ritchie, D.A., (1975), *Molecular Virology*. Chapman and Hall, with permission).

2.3 The genomes of bacteria, their structure and regulation

Bacterial genomes are enormous by viral standards but small in comparison to those of eukaryotic organisms. All have double stranded DNA genomes in the form of closed circles, that of *E.coli* being about 1100 μm long, consisting of approximately 3.4×10^6 base pairs and including coding sequences for about 1800 genes. The DNA is not complexed with histone to form chromatin, but its negative charge is neutralised by association with small polyanions. Some other prokaryotic organisms, such as *Mycoplasma*, *Chlamydia* and *Rickettsia*, have genomes which are only about a third of the size of E. coli, and *Bacillus megatherium* has a DNA amount more than five times that of *E. coli* and some spirochaetes have even more.

Hundreds of genes have been assigned to specific loci on the *E. coli* genome, and their distribution and arrangement reveal some interesting aspects. One feature is that genes involved in the same metabolic pathway show a marked tendency to be clustered, and this topic will be discussed further in Chapter 4 in the context of operons. Another noticeable feature is that some regions of the genome have very few genes assigned to them, and other equivalent lengths have scores of coding sequences alloted. Now, in part, this may result from the non-random nature of the genes chosen, selected largely because their products are known and mutants have been studied by recombinational analysis. But it is hard to believe that this is the only explanation, and it is more likely that some tracts of the genome are indeed made up of rather silent stretches of DNA with few coding sequences. It is also noticeable that many genes, especially those which code for proteins which are required in large amounts by the cell, are located near to the origin of DNA replication. This is because there is a single point within the circular genome of a bacterium at which DNA replication begins, proceeding out in both directions to end at a termination point on the other side of the circle. Often in a bacterial cell, growing in rich media, a number of rounds of DNA replication will initiate before the first has time to be completed. This has the effect of increasing the number of copies of genes placed near to the origin, as compared with those which are distant. So clustering near to the replication origin is no doubt a result of evolutionary selection in terms of demand for the gene product.

2.3.1 Bacterial genomes often contain transposable elements

A class of mutation has been known in bacteria for some time that betrays its presence by preventing expression of a gene or genes but cannot be cured by simple addition of bases or frame shift mutations; in other words such polar mutations could only result from the insertion of a substantial piece of DNA into the promoter sequence upstream from the affected coding sequences. But where has the inserted DNA come from? It was gradually realised that the bacterial genome contained numbers of highly mobile genetic elements, each characterised by the presence of numerous stop signals and chain termination mutations (which account for the polar mutations resulting from their

Properties of some *E. coli* insertion elements

Element	*Number of copies and location*	*Size, base pairs*	*Sequence data*
IS1	5-8 in chromosome	768	Complete
IS2	5 in chromosome, 1 in F	1327	Complete
IS3	5 in chromosome, 2 in F	Approx. 1400	Nearly complete
IS4	1 or 2 in chromosome	Approx. 1400	Termini only
IS5	Unknown	1250	Termini only
γδ	1 or more in chromosome, 1 in F	5700	Termini only

Fig. 2.12 Characteristics of different insertion elements found in the genome of *E. coli*. (From D. Freifelder, *Molecular Biology*, Van Nostrand Reinhold with permission).

insertion), and by having inverted repeat sequences at their termini. These were termed IS elements (insertion elements) and they constitute one of a number of different types of transposable sequences (*transposon*) found in the bacterial genome. As seen in Fig. 2.12 there are a number of different IS elements in *E. coli*. Each IS element carries at least one gene, coding for a protein necessary for transposition, but some have two genes. IS elements are the simplest transposable elements, and a somewhat more complex type, known as composite type 1 transposons, often consist of a gene or genes for antibiotic resistance flanked by simple IS elements at either end. Other large types of transposons may be 5000 base pairs or more in length, containing a series of genes or even an entire phage. All are flanked by repeat IS elements, and it is believed that the formation of a short stem of hydrogen bonded base paired DNA between the terminal repeats facilitates the integration of the element into the bacterial genome. Many plasmids within bacterial cells contain integrated transposons, as do a number of phage, and these elements clearly enable the plasmid or phage to insert or to exit from the bacterial genome rather freely. Insertion sites for possible transposition are widely scattered in the bacterial genome, but some of the small IS elements have a very restricted number of possible sites. As we shall discuss further in Chapter 6, transposable elements are not confined to the genomes of prokaryotes as some can be found in eukaryotes.

2.3.2 Prokaryotic transcription is controlled by promoter sequences

Unlike eukaryotes, prokaryotes have a single species of RNA polymerase enzyme, the protein which is responsible for transcribing genes into messenger RNA. Since one enzyme is responsible for transcribing all genes, it is obvious that the specificity of the transcription, that is the decision to transcribe only certain genes at certain times, must be regulated by factors other than the enzyme itself. This RNA polymerase enzyme is a multi-subunit molecule, three of these subunits making the so called core enzyme. In cell free systems, when the core enzyme is provided with a DNA template, RNA synthesis is initiated indiscriminately throughout the DNA available. A further subunit, termed sigma factor, which is commonly part of the molecular complex within the bacterium, confers specificity on the enzyme complex, and it then shows marked preference for promoter sequences (which, as in eukaryotes, lie upstream of the coding regions of the DNA). The promoter sequences themselves consist of two regions; one, known as the Pribnow box, consists of the consensus sequence TATAAT and lies ten bases upstream from the site at which the coding sequence region commences. A second promoter often simply termed the recognition sequence, lies some 20 bases further upstream and consists of the sequence TTGACA. The number of bases between these two regions is optimally 17, and apparently the identity of these intervening bases is unimportant. What has been described above are the *E. coli* promoter sequences: those of other bacterial species show some slight variation. However, all bacterial genes carry such promoters, and presumably interplay between sigma factors and other transcription factors is responsible for determining which genes are transcribed at particular periods within the cell cycle. So the promoters ensure that transcription starts in the right places and that RNA polymerase enzymes are available to do the job, while other factors, discussed more fully in Chapter 4, determine which genes are expressed at what times.

2.3.3 A specific sequence act as a replication origin

The bacterial 'chromosome' functions as a single replicon, with one point of initiation from which the DNA replication molecules move out in either direction to form two replication forks. These forks eventually meet up at the point of termination when replication is completed. One of the reasons why it has always proved difficult to make precise determination of the size of the bacterial haploid genome by colorimetric assays of DNA is that the chromosome was often in the process of replication and may even have more than one round of replication ongoing simultaneously. Thus erroneously high figures for DNA content were obtained from cultures containing a significant proportion of such cells. The replication origin consists of a sequence of about 1000 nucleotides, within which a sequence of 440 base pairs is essential for bidirectional replication to proceed. Several short consensus sequences have been discovered in this region by comparing sequences in different bacteria. Many of approximately 12 base pairs in length are found to alternate between spacers of some 16 base pairs, and it is assumed that the various proteins of the replication machinery recognise the short 12 base pair sequences.

2.3.4 Gene arrangement within the bacterial chromosome

We have already established that bacterial genes are often clustered within the circular chromosome. It is also interesting to note that there is no highly repetitious DNA in prokaryotes, nor are there any intron sequences within the genes. There are, however, some moderately repetitious sequences. These will include the consensus sequences within promoters already alluded to. Although the minute prokaryotic cell *Mycoplasma* has only single copies of the ribosomal genes coding for 16s and 23s rRNA, *E. coli* has seven of each, and as in eukaryotes, these are linked and a polycistronic transcriptional unit is made combining the 16s and 23s genes plus 5s and some tRNA genes. Multiple genes for other tRNAs are also present in *E. coli*, but are located elswhere in the genome.

2.3.5 Prokaryotes have developed some special mechanisms to help increase genetic diversity

As with eukaryotic DNA, mutational events and error-prone repair mechanisms in bacteria constitute the basis of evolutionary change and the spread of genetic diversity. But other mechanisms help to ensure that mixing of DNA occurs, thus permitting novel gene assortments to arise.

There are three different mechanisms involved here, and space does not permit more than a brief account of them. The first is *transformation*, initially discovered by Griffiths in 1928, and involving spontaneous uptake of DNA by a living bacterial cell from the surrounding medium, and eventual recombination of some of these molecules with sequences in the bacterial chromosome. This process permits living bacteria to acquire genetic properties spontaneously from the surviving DNA of recently dead bacteria. A second and much more complex mechanism is *conjugation*, the transfer of a sex plasmid from a 'male' to a 'female' bacterial cell via a sex pilus, (plasmids will be discussed as a separate topic in a following section of this chapter), or, in the case of HFr conjugation, the transfer of a section of the chromosome from one bacterial cell to another of a different mating type. In this case the sex factor itself is rarely, if ever, transferred. The recipient bacterium may then experience recombination between the incoming DNA sequence and its own chromosome. It was quickly found by experimentalists that conjugation between HFr strains could be readily terminated by shaking, and by such interrupted mating experiments between differing strains, the order of genes on the bacterial chromosome could be determined and the circularity of the entire structure made evident (this in fact was how it was discovered). Fig. 2.13 illustrates the mechanisms of these two variant forms of bacterial conjugation.

The third mechanism, *transduction*, is essentially phage assisted transformation, since it involves the transport of a sequence from the bacterial genome of one cell into another via a defective phage. When a virulent phage, such as lambda in its lytic phase, is replicating within a bacterial host cell, an accidental

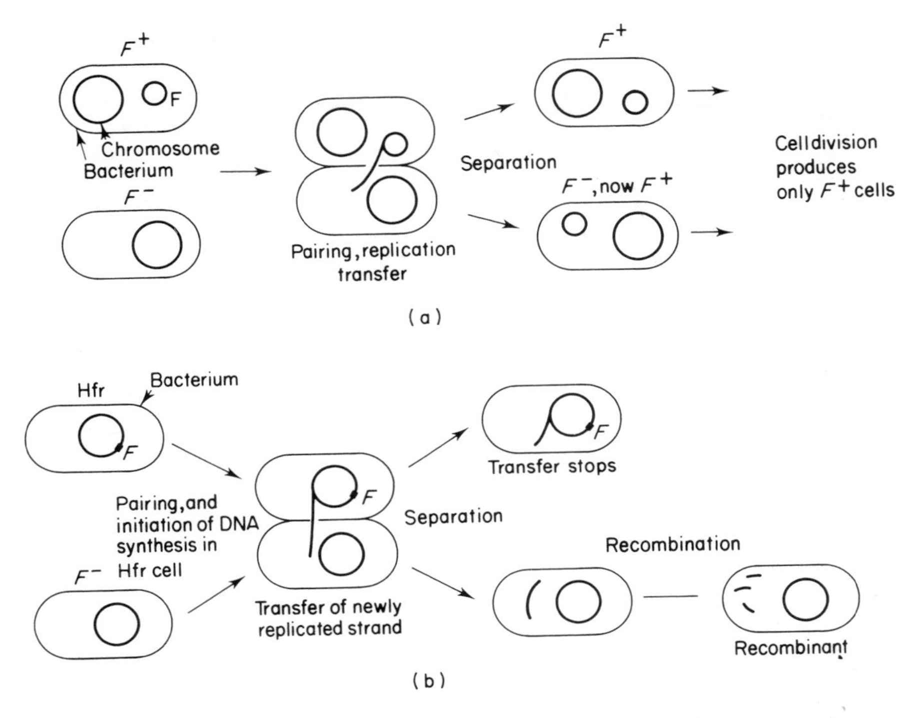

Fig. 2.13 Diagrams illustrating the two main types of conjugation found in *E. coli*. In (a) the effects of simple mating between F+ and F− bacteria is shown, while in (b) the effects of mating between an HFr male and an F− female are shown. The female cell is not normally converted to male in this process but some of the transferred bacterial DNA fragments may recombine with sequences in the chromosome, resulting in a female cell with new genetic properties acquired from the male. (From D. Freifelder, *Molecular Biology*, Van Nostrand Reinhold, with permission).

recombination event occurs between a piece of phage DNA and a piece of the host cell genome. When this hybrid molecule is packaged in the virion it will then be released and may infect a second cell. On infection the phage will not lyse the new host since it is itself genetically defective as a result of acquiring the bacterial DNA, but this DNA may now in turn be exchanged with a part of the genome of the new host, thus completing the transduction event.

2.3.6 Bacterial plasmids

Plasmids are mini circles of double stranded DNA found in bacteria (and also in some eukaryotic cells), generally quite separate from the main bacterial chromosome, and capable of independent replication. They are composed of DNA of between 10^3 and 10^5 base pairs and are inherited by daughter cells at cell division. The numbers of plasmids per cell varies greatly and can be less than ten per cell. When the number is small, say less than ten, they are said to be controlled *stringently*. In the alternative state, known as *relaxed* control, plasmid numbers may reach 50 per plasmid type per cell. Yields of up to 1000 plasmids per cell may build up if bacteria are grown in the presence of chloramphenicol, and this trick is widely used in harvesting plasmids involved in recombinant DNA technology.

Plasmids carry genes which ensure their own survival, and have an independent replication origin recognised by the bacterial polymerase enzymes. Three distinct classes of plasmids have been recognised in bacterial cells. Firstly, *R factors*, which are plasmids carrying genes for antibiotic resistance (these genes produce products which work in different ways, either by direct detoxification of the antibiotic, or by interfering with antibiotic uptake). Secondly, there are *col factors* producing colicins (in the case of *E. coli*), or, more generally, bacteriocins – proteins that are toxic to other strains of bacteria not carrying the factor. The third class involves the sex factor plasmids, such as the F factor in *E. coli*, which mediates the conjugation process.

Since plasmids can be harvested in large numbers and at high purity, and can be readily manipulated by cutting with specific restriction endonuclease enzymes, they have opened the way for widespread exploitation in genetic engineering. In such work, plasmids are purified, dissected by enzymic digestion, and spliced to some other DNA sequences which are to be cloned. The circular plasmid is then closed by enzymic manipulation and ligation, and bacteria infected with the new chimeric plasmids. Such infection does not require a specific mechanism, such as conjugation, but is achieved by *transfection*, a process in which spontaneous uptake of novel DNA is accomplished by cells incubated in media containing the plasmid DNA and also rich in calcium phosphate or calcium chloride. Following successful transfection the cells containing the novel plasmid are grown on, then incubated in the presence of chloramphenicol, thus permitting the recovery by cell lysis and centrifugation of a high yield of millions of copies of the specific chimeric plasmid.

2.3.7 The genomes of mitochondria and chloroplasts

One of the astonishing aspects of eukaryotic cell organisation is that all such cells contain a series of partially autonomous and self replicating genetic systems. It is likely that both mitochondria and chloroplasts are evolved from prokaryotic cells that lived more or less symbiotically with ancestral eukaryotic cells. Indeed some protozoan cells such as those of *Paramecium*, continue to live with commensal intracellular bacteria or green alga. The former occur in certain strains of *Paramecium aurelia* and are known as kappa particles, and the latter in the green *Paramecium bursaria*.

Since there are radical differences between mitochondria and chloroplasts, the account that follows will begin by listing the way in which both particles are distinctive and their common genetic properties, and this will be followed by a

Table 2.3 Contribution of mitochondria and chloroplasts to the cell and to cellular DNA contents. Notice that amphibian eggs have a mitochondrial cloud, a large stockpile of mitochondria which are distributed amongst the embryonic cells by division. Complexity indicates the combined total length of all differing sequences as determined by reassociation kinetics. (From Borst, P., *et al. Extranuclear Genes in Eukaryotic Genes*, N. Maclean, *et al.* (ed), Butterworths, 1983, with permission).

Source of DNA	*Organelles per cell*	*Organelle DNA (% of total)*	
		Amount	*Complexity*
Mitochondrial DNA			
Mouse L-cell line	10^2	1	0.0005
Toad egg	10^7	99	0.0005
Yeast diploid	2–50[(a)]	15	0.6
Chloroplast DNA			
Chlamydomonas diploid	2	7–14	0.3
Tobacco leaves	10^2	10–50[(b)]	0.01

[(a)]Depends on growth conditions.
[(b)]Controversial.

brief separate consideration of the two particles. As seen in Table 2.3, mitochondrial DNA accounts for between 1 and 15% of the cellular DNA, except for the staggering 99% of cellular DNA contributed by the mitochondrial cloud of some vertebrate eggs. Chloroplasts account for between 7 and 20% of the total cellular DNA of plant cells. All chloroplast genomes and almost all mitochondrial genomes are closed circles of double stranded DNA. The exceptions are the linear DNA genomes of some protozoan mitochondria, and the huge concentrated network of connected circles which form the genome of the giant mitochondria found in some species of trypanosome. Such DNA, whether circular or linear, is not in the form of chromatin, since no histones are present, and so the genomic DNA of these organelles closely resembles that of bacteria.

Chloroplast genomes are substantially larger than those of mitochondria although the number of organelles per cell may be roughly similar, say about 100 for a cell of average size. In both, introns are rare or absent in the genes, and the total numbers of genes present is quite small, less than 50 in mitochondria and less than 100 in chloroplasts, and of these between 20 and 40 in either case code for tRNA. What has clearly happened in both cases, but especially so with mitochondria, is that genes have migrated from the mitochondrial to the nuclear genome and so, although no copy of the mitochondrial genome is present in the nucleus, mitochondria are now heavily dependent on nuclear genes for most of their own proteins. A brief inventory of known mitochondrial and chloroplast genes is given in Table 2.4. Another curiosity of mitochondrial and chloroplast genetics is that both seem to be invariably of purely maternal origin, implying that the mitochondrion which provides the energy for sperm motility is destroyed and makes no contribution to the embryonic mitochondrial store. This has been confirmed by studying the mitochondrial DNA of hybrids

Table 2.4 A brief inventory of known mitochondrial and chloroplast genes.

A An inventory of mitochondrial genes in yeast and man

Mitochondrial component	Mitochondrial gene product in	
	Yeast [a] (a/α)	Man [b] (♀/♂)
Cytochrome c oxidase		
Subunit I	+	+
Subunit II	+	+
Subunit III	+	+
Ubiquinol-cytochrome c reductase		
Apocytochrome *b*	+	+
ATPase complex		
Subunit 6	+	+
Subunit 9	+	–
Large ribosomal subunit		
rRNA	+	+
Small ribosomal subunit		
rRNA	+	+
Ribosome-associated protein	+	– ?
RNA processing enzymes		
Intron 2 COB maturase	+	–
Intron 4 COB maturase	+	–
tRNAs	about 25	22
URFs	⩾8	8

B Genes in chloroplast DNA

Structural RNA genes
35–40 tRNAS.
1–3 sets of rRNA

Genes for ribosomal proteins
At least one.

Genes for membrane proteins
ATPase complex: 3 subunits (out of 5) of the CF_1 part. One subunit (out of 3) of the CF_0 part.
'Photogene 32' (= gene for a 32 kD protein of the thylakoid membrane, which is made in response to light).
Cytochrome *f*.
Cytochrome b_{599}.

Genes for stromal proteins
Large subunit of ribulose-bisphosphate carboxylase.
Elongation factors G and T[a]

Reproduced, with permission, from Borst, P. in 'Eukaryotic Genes' N. Maclean *et al.* (edit) Butterworths 1983 between species crossed in both ways. For example, mules have been shown to have horse mitochondrial DNA and hinnies donkey mitochondrial DNA – the mule is a male donkey × female horse and the hinny is a female donkey × male horse.

Some special features of mitochondrial genomes

Some special features of interest are to be found in the structure and function of mitochondrial genomes. For example, as shown in Table 2.5 the genetic triplet code which applies so ubiquitously in all living systems is slightly altered in mitochondria, and varies slightly between mitochondria of different species. Thus the UGA triplet with functions as a stop codon in all prokaryotic and eukaryotic cells codes for the amino acid tryptophan in all mitochondria, while AGA, which codes for arginine in the mRNA of all cells, is a stop codon in the mRNA of human mitochondria.

Another point of interest is the existence of maturase enzymes in mitochondria, and maturase genes in the mitochondrial genome. These proteins are involved in the splicing of mRNA precursors in mitochondria. Some mitochondrial genes, for example, those coding for cytochrome enzyme subunits in the mitochondria of yeast, are interrupted by introns, and, remarkably, the genes for maturase are located within the introns of apocytochrome genes of yeast mitochondria. The presence of introns in cells such as yeast has lent support to the view that introns may be a primitive rather than an advanced feature in cellular evolution and that present-day prokaryotes may have lost introns that were present within the genes of their ancestors.

Most of the statements made about mitochondria are true for both plant and

Table 2.5 Assignments of coding triplets in message to particular amino acids (or stop signals) in eukaryotic or prokaryotic genomes (usual assignment), and in mitochondria of three different organisms. Thus UGA is a universal stop signal, but codes for tryptophan in all three mitochondria. (From Borst, P., *et al.*, *Extranuclear genes in Eukaryotic Genes*, N. Maclean, *et al.* (ed), Butterworths, 1983, with permission).

		Assignment in mitochondria of		
Codon	*Usual assignment*	*Man*	*Neurospora*	*Yeast*
UGA	Stop	Trp	Trp	Trp
(UGG	Trp	Trp	Trp	Trp)
CUA	Leu	Leu	Leu	Thr
(ACX	Thr	Thr	Thr	Thr)
AUA	Ile	Met	Ile	Ile
(AUG	Met	Met	Met	Met)
AGA	Arg	Stop	Arg	Arg
AGG	Arg	Stop	Arg	Arg
(CGN	Arg	Arg	Arg	Arg)

animal cell mitochondria, but some particular comment concerning plant mitochondria is appropriate. Firstly, they frequently contain very large genomes, often ten or even 100 times larger than those of say, human cell mitochondria. Secondly, some of these very large genomes coexist with smaller ones, and are made up of a series of repeated elements, some of which appear to have arisen from others by recombination events. The manner in which such recombination is engineered is still unclear (Fox, 1984) but a general system of recombination has been discovered in relation to the mitochondrial DNA of yeast. Most of the mitochondrial genome in plants does not seem to code for useful products, although copies of chloroplast genes have been discovered within the genomes of maize mitochondria. These organelles seem to be much more efficient at gaining rather than losing DNA.

Some special features of chloroplast genomes

As has been stated earlier, chloroplast genomes are substantially larger than those of mitochondria and carry many more genes, only a few of which have been identified. Fig. 2.14 shows a simple diagram of a chloroplast genome with known gene assignments. The most remarkable feature about the chloroplast

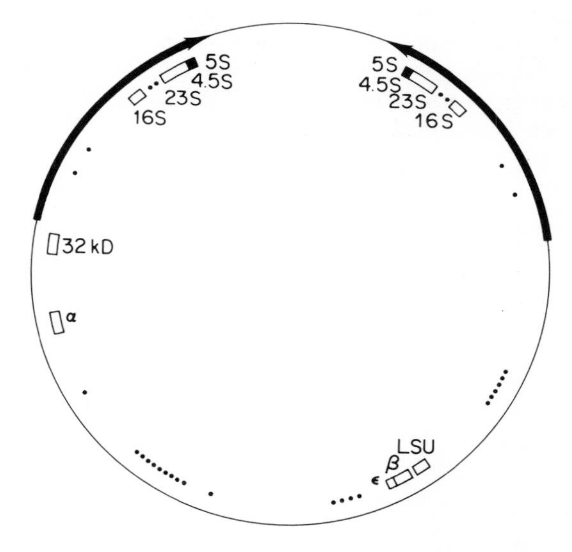

Fig. 2.14 Partial map of gene organization in chloroplast DNA of higher plants. The two thickened lines terminating in arrows are the long terminal repeats, black dots are genes coding for tRNA, and the ribosomal genes are designated as 16S, 23S, 4.5S, and 5S. Genes coding for subunits of the ATP ase complex are donated by α, B and E, the gene for the large subunit of ribulose – 1, 5 bisphosphate carboxylase by LSU, and the light induced protein of 32KD has its gene designated as 32KD. (From Borst, P., *et al. Extranuclear Genes in Eukaryotic Genes*, N. Maclean, *et al.* (ed), Butterworths, 1983, with permission).

genome is its similarity to that of bacteria such as *E. coli*. For example, the genomic organisation of the large ribosomal genes is strikingly similar between chloroplast DNA and *E. coli* DNA and the 16s RNA shows 74% homology between *E. coli* and maize mitochondria. In spite of this, some chloroplast genes possess introns, for example, in two of the tRNA genes of maize chloroplast DNA. Why only some chloroplast genes have retained introns is as yet unclear. Also, chloroplast promotion is very similar to that of *E. coli* and indeed the promotors are recognised by the *E. coli* RNA polymerase enzyme.

2.4 Concluding remarks

This chapter has attempted a very brief review of the organisation of small genomes. The diversity of structure and function is astonishing and the basic economy of the genetic code is stretched to its limits in some of these situations. The problems that are encountered are the opposite of those apparent in the genetic system of high eukaryotes. In these small genomes it is the economy of use that is so striking, and the great efficiency with which the same DNA may be pressed into service to provide information for a number of different proteins. In higher cell genetics the amazing and contrasting situation is that organisms seem to live efficiently with gigantic genomes, accumulated through the elaborate evolution of higher animals and plants, and apparently containing much information that is seemingly quite useless to the organism concerned. Here the wonder is why there is so little selective pressure for eliminating the apparently redundant material.

3

Chromatin in Eukaryotes

A startling complication arises when we move from the prokaryotic genomes discussed in the previous chapter to consider the much larger genomes of eukaryotes, namely that the DNA is now conjugated with protein to form chromatin. Eukaryotic cells also have nuclei, of course, and so somewhere in the mists of early evolution there arose this remarkable development of placing all the genomic DNA within a nuclear envelope and packaging it into chromatin. Without exception all eukaryotic cells display these two features, features which are quite unknown amongst prokaryotes.

Chromatin consists at a rough approximation of a 50/50 mix of DNA and protein. More precisely it is about 35% DNA, 35% basic protein, 10% RNA and 20% acidic protein. Since the basic protein is, in almost all cases, histone, the acidic protein fraction is often referred to simply as non-histone protein. A fuller account of chromatin structure will be found in Bradbury *et al.* (1981).

3.1 Chromosomal proteins

3.1.1 Histones

In all eukaryotic cells, except sperm, the DNA is complexed with one group of proteins, the histones. In sperm, where very great compaction of chromatin is required, histone is replaced by another basic protein called protamine. Histones are small proteins, rich in the amino acids lysine and arginine, which confers on them their net positive charge at neutral pH. This positive charge helps to neutralise the net negative charge of the DNA molecules. In most somatic tissues histones extracted from chromatin can be resolved into five different classes of protein, termed H1, H2A, H2B, H3 and H4. In some cells such as the erythrocytes of birds, amphibians and fish, another histone, H5, occurs, and in these cells it supplements or replaces histone H1. Histones are readily separated by polyacrylamide gel electrophoresis, since their comparatively low molecular weight allows their easy separation from the bulk of cellular proteins (see Fig. 3.1). Other small acidic proteins migrate just behind

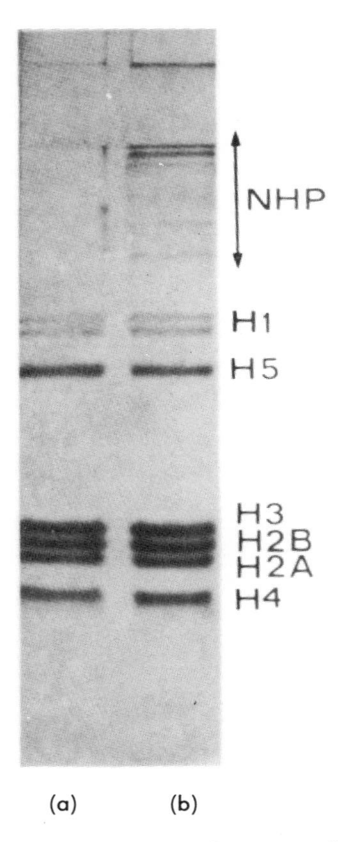

Fig. 3.1 Polyacrylamide gel patterns of histones from (a) chicken erythrocyte chromatin and (b) chicken erythrocyte nuclei. NHP marks the non-histone proteins recovered, some of which will be HMG (high mobility group) proteins. The two bands of HI indicate the presence of two slightly variant forms of HI, and H3, H2B, H2A and H4 are the core histones present in equimolar amounts. H5 is an HI-like histone which partially replaces HI in the chromatin of nucleated erythrocytes, where chromatin is very highly compacted. (From Bradbury, E.M., *et al.*, *DNA, Chromatin and Chromosomes*, Blackwells (1981) with permission).

the histones, and these non histone chromosomal proteins have come to be known as high mobility group protein (HMG proteins).

Histones are remarkable molecules in a number of ways, one being that they have been very closely conserved during evolution. Thus histones H3 and H4 are the most conserved of all proteins so far studied in that in a molecule of over 100 amino acids, only two amino acid replacements have occurred in H4 in the course of evolution between the pea plant (*Pisum sativum*) and the ox (*Bos taurus*). Although it is believed that the lines of evolution of plants and animals diverged over one thousand million years ago, their H4 sequences are virtually identical. Bovine and pea H3 are separated by 4 amino acid substitutions and the histones H2A and H2B are only slightly less conserved, for example, six changes between trout and ox in H2A. What can we deduce from this remarkable conservation? There are two conclusions which can be drawn. One is that these proteins must serve a crucial cellular function, and the other that almost their entire primary sequence of amino acids must be essentially involved in the function, so that any mutational change in amino acid sequence would be strongly disadvantageous and so heavily selected against. As we shall see, both of these conclusions hold true in the context of the known functional interactions between histones and the DNA.

3.1.2 Histones and Nucleosomes

The conformational relationship between DNA and histone remained a poorly understood area of investigation for many years. Studies by both light and electron microscopy revealed little of the wanted details of structure when chromosomes were scrutinised, but it was widely assumed that the histone molecules must lie within the grooves of the DNA helix, or at least be involved in an interaction in which the DNA was on the inside and the histone on the outside. A glance at any genetics or cell biology text written before 1970 will reveal a range of proposed models for chromatin all based on the assumption that the histone fits around the DNA. However, in the early 1970s two Australian biochemists, Hewish and Burgoyne, carried out a most revealing experiment. They made a homogenate from rat liver tissue, provided added calcium ions to permit the activation of calcium-dependent nuclease enzymes, and ran the ensuing DNA fragments on an electrophoretic gel. To their amazement, a regular stepladder of DNA fragments was visualised, the smallest being of a single band equivalent to 200 base pairs, the next of 400, the next 600, the next 800 and so on. If digestion was allowed to proceed to completion, all the DNA was reduced to a fragment length of 200 base pairs. Figure 3.2 illustrates the result of a similar experiment using an exogenous nuclease. What this experiment indicated was that at intervals of approximately 200 base pairs the DNA was particularly sensitive to enzymic digestion, either because of some peculiarity of the DNA, or because it was protected in some way in all the non-digested areas. Since there was no evidence that DNA differed in any structural way at 200- base pair intervals, it seemed that there was here compelling evidence for partial protection by some other molecule, and most probably histone.

At this time some other evidence about chromatin structure was also coming from the electron microscopy front. Those skilled with this instrument had fought long and hard to make something of chromatin fine structure, and now pictures appeared of chromatin spread on films, indicating a beads-on-a-string arrangement. Excitingly the beads, of 11–12 nm diameter, were spaced on the DNA string at intervals of about 200 base pairs! Putting these two observations together, the concept of the nucleosome was born. A nucleosome was conceived as a histone DNA complex, containing eight molecules of histone and associated with up to 200 base pairs length of DNA. Studies by nuclear magnetic

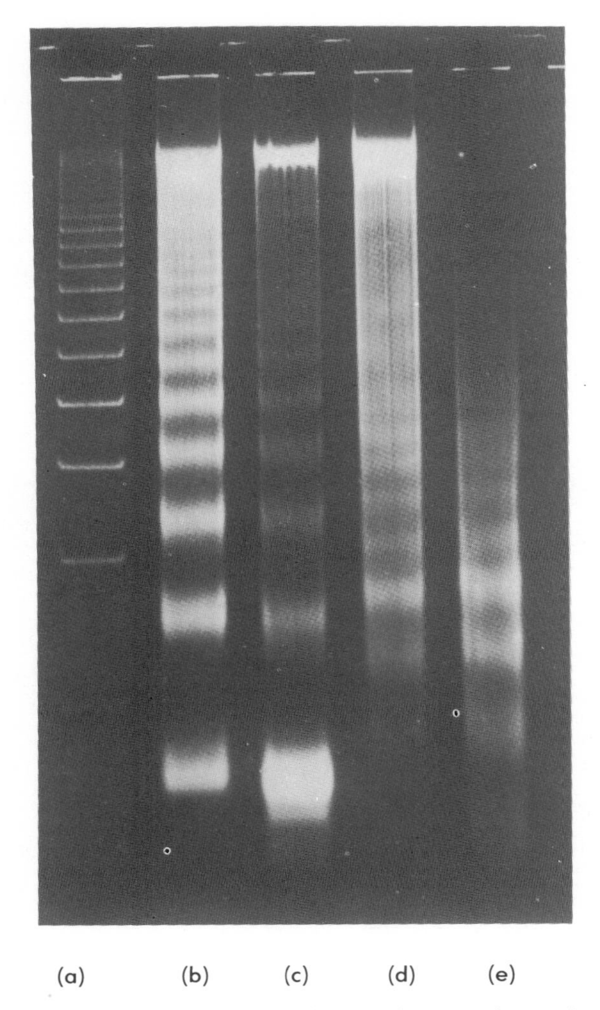

Fig. 3.2 An experiment run on similar lines to the original Hewish and Burgoyne (1973) experiments, except that the endogenous calcium – activated endonuclease has been replaced by micrococeal nuclease or DNase II. Chromatin has been isolated from mouse liver nuclei and incubated for 1 hour at 37°C with micrococcal nuclease (lanes (b) and (c)) and DNase II (lanes (d) and (e)). The concentrations used were 7 and 25 units/ml in (b) and (c) and 125 and 250 units/ml in (d) and (c). Lane (a) contains a mouse satellite DNA marker after digestion with Eco RIII. The ladder of fragments has a periodicity of 200, 400, 600 . . . base pairs of DNA running from bottom to top, while lane (c) contains one main band of 200 base pairs and traces of 400 and 600 base pair bands. The DNase II ladder is of 100 base pair intervals and involves cutting the DNA within the nucleosome bead. (From Altenburger, F.R.M., *et al.*, *Nature*, **264**, 517 (1976) with permission).

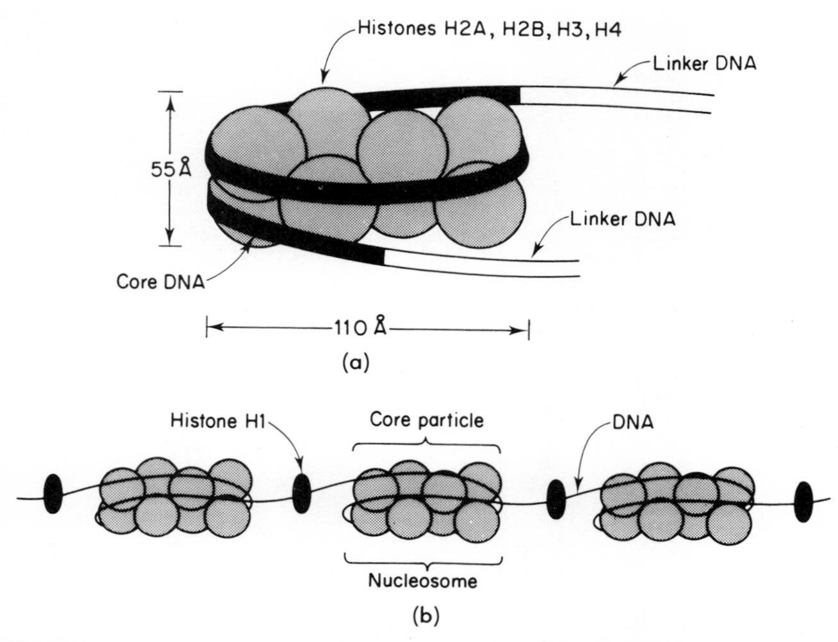

Fig. 3.3 (a) A Suggested structure for the nucleosome, showing the octamer of histones in the core, the double loop of DNA round the outside, and the linker DNA. (b) A suggested structure for chromatin showing a series of three nucleosomes as 'beads-on-string', with the HI histone associating with the linker DNA. (From D. Freifelder, Molecular Biology, Van Nostrand Reinhold, with permission).

resonance techniques soon indicated that it was likely that the histone was on the inside of the bead and the DNA wound around the outside.

Present day understanding of nucleosome organisation, as seen in Fig 3.3, is of two molecules of H2A, H2B, H3 and H4, with approximately 140 base pairs of DNA looped in two circles of 70 base pairs around the histone core. A tetramer of H3 and H4 forms the inner part of the core, with an H2A H2B dimer taking up an outer position on either flank. The spare 30–40 bases length of DNA forms the linker region between nucleosomes. The fifth histone, H1, seems to be involved in forming a rather looser association between the nucleosome core and the linker DNA, and thus has a role in condensing or relaxing the whole beads-on-a-string chromatin 'necklace'. Some further evidence also indicates that the DNA is not looped smoothly around the core, but is slightly kinked at intervals of approximately 10.4 bases. Another nuclease enzyme, DNase I, will cut DNA at these kinks and yield a ladder of DNA fragments from chromatin digestion each fragment being approximately 10 base pairs in length (see Fig. 3.2 (d) and (e)).

How then is this string of nucleosomes on the DNA thread packaged into the structures that we know as chromosomes? The first order packaging of the

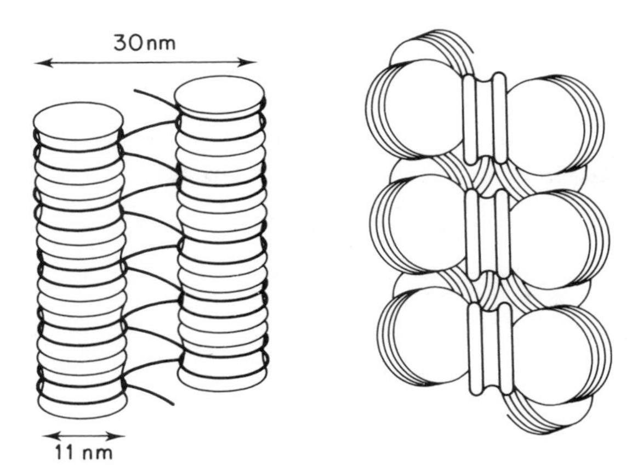

Fig. 3.4 Two alternative ways in which the solenoidal array of nucleosomes in the basic 30 nm chromatin fibre may be organised.

nucleosome beads on the DNA string is fairly clear, namely that the nucleosomes adopt a solenoidal compaction pattern, as shown in Fig. 3.4, to make a chromatin fibre of diameter 34 nm, usually known as the 30 nm fibre. Although the precise packing arrangement in the fibre is still not completely understood, the existence of the 30 nm fibre of compacted nucleosomes is not in any doubt. There appears to be a transition from condensed to decondensed chromatin depending on such factors as gene activity and this transition clearly involves the activity of H1 histone. What happens to chromatin in situations of gene activity and genetic shutdown is a topic which we will return to shortly. It is a considerable leap from a chromatin fibre of 30 nm to a mitotic chromosome with a diameter of 700 nm, and the details of this further aspect of higher order chromatin structure is as yet poorly understood. The most probable arrangement is that a central chromosome axis, endowed with non-histone proteins which help to form a basic scaffold, provides some rigidity and anchorage for long loops of the 30 nm chromatin fibres which are subtended by the scaffold. The electron microscope pictures of chromosome spreads which have been depleted of histone certainly confirm this view of the arrangement of chromatin arrangement within the chromosomes (See Fig. 3.5), and is further supported by our understanding of the structure of lampbrush chromosomes, discussed later in this chapter.

Of course it must be remembered that interphase chromatin is not in the form of heavily compacted chromosomes, (with the exception of particular bits such as the inactivated X chromosome). Nevertheless, it is clear that even in the interphase nucleus, when discrete chromosomes are not visible, the chromatin remains substantially condensed and retains its general disposition into the separate chromatin blocks.

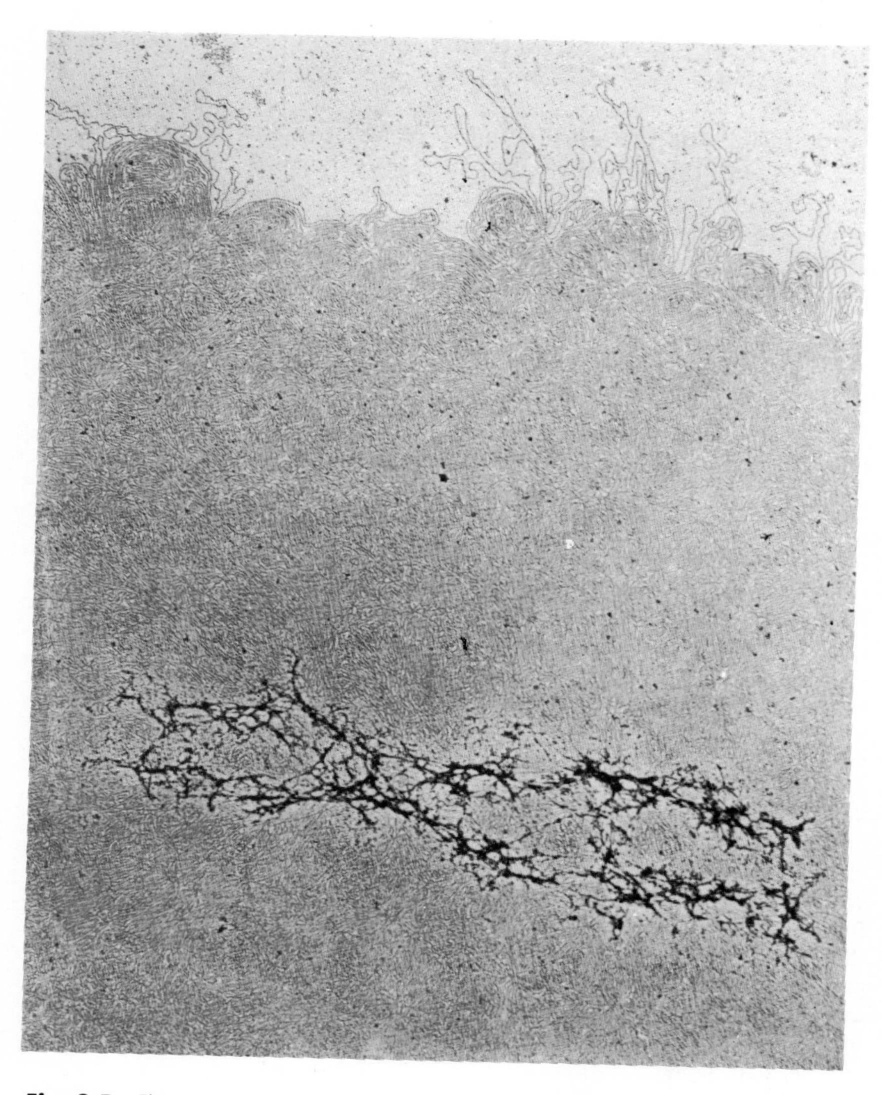

Fig. 3.5 Electron micrograph of part of a metaphase chromosome from a human cell, following histone removal. The dark material is a central axis containing matrix of non-histone chromosomal proteins, and the DNA loops extending out from the matrix area are of the order of 30 microns in length. No free ends of DNA are visible. (From Paulson, J.A. and Laemmli, U.K., *Cell*, **12**, 817 (1977), with permission).

Active and inactive chromatin

As soon as the existence and structure of nucleosomes became known, the question arose as to whether the nucleosomal organization of chromatin persisted in confined regions in which genes were being actively transcribed. After all, the nucleosome is likely to pose a considerable hurdle to the RNA polymerase enzymes as they traverse the DNA, itself partially opened up to allow the code to be copied into RNA. Since RNA polymerase molecules are only slightly smaller than nucleosomes, the obstacle course might seem almost insuperable. As yet this problem has not been resolved. Some active DNA regions, especially those in the vicinity of actively transcribing ribosomal genes, do seem to be nucleosome free, as revealed by electron microscopy. But in other situations there seems to be equally good evidence for the persistance of histone on active gene regions. Perhaps in the latter case the nucleosomes simply change shape to offer less resistance to the transcriptional machinery.

It is quite clear that transcriptionally active chromatin is usually in an extended open conformation as compared with inactive chromatin, and so the condensed 30 nm fibres probably represent chromatin containing genes which are transcriptionally inactive. The evidence comes from experiments in which chromatin is exposed to the enzyme DNase I. By the use of this procedure it has become evident that this enzyme will preferentially cut chromatin containing active genes. Using radio tagged gene probes, it can be readily shown that DNase I – digested chromatin lacks genes known to be active in the tissue of origin, but retains genes known to be inactive in such a tissue. Clearly the chromatin condensation in the latter regions protects the genes from attack by the enzyme.

3.1.2 Non histone proteins

A wide range of proteins are associated with chromatin at different times in the life of a cell, but many of these are only transiently associated. Thus RNA polymerases, proteins involved in DNA replication, and the specific gene regulatory proteins to be discussed in Chapter 4, are all acidic proteins with some degree of chromosomal association at localised sites and times. But the HMG proteins mentioned in section 3.1.1 do seem to have a more permanent association and some, known as HMG 14 and 17, are suspected of having a structural role in chromatin. The proteins which contribute to the chromosomal scaffold on which the DNA loops are supported must also have a fairly permanent relationship with chromatin, but their identity remains to be unambiguously demonstrated.

3.1.3 Protamines in sperm

In the sperm of some, but not all, animals, histone is entirely replaced by an alternative basic protein known as protamine. Many fish have protamine in the sperm cells, since these are somewhat variable forms of protamine they have been given distinct names – salmine in salmon, iridine in trout, thynine in

Fish protamines

CLUPEINE Y1	Ala	Arg Arg Arg Arg	Ser Ser Ser Arg Pro Ile	Arg Arg Arg Arg	Pro Arg Arg Arg	Thr Thr	Arg Arg Arg Arg	Ala Gly	Arg Arg A
SALMINE A1	Pro	Arg Arg Arg Arg	Ser Ser Ser Arg Pro Val	Arg Arg Arg Arg Arg	Pro Arg	Val Ser	Arg Arg Arg Arg Arg Arg	Gly Gly	Arg Arg A
IRIDINE I1	Pro	Arg Arg Arg Arg	Ser Ser er Arg Pro Val	Arg Arg Arg Arg	Ala Arg Arg	Val Ser	Arg Arg Arg Arg Arg Arg	Gly Gly	Arg Arg Ar
THYNINE Y1	Pro	Arg Arg Arg Arg	Glu Ala Ser Arg Pro Val	Arg Arg Arg Arg Arg	Tyr Arg Arg Ser Thr	Ala Ala	Arg Arg Arg Arg Arg	Val Val	Arg Arg Ar

Bull protamine

Ala Arg Tyr Arg Cys Cys Leu Thr His Ser Gly Ser Arg Cys Arg Arg Arg Arg Arg Arg Arg Cys Arg Arg Arg Arg Arg Arg Pro Gly Arg Arg Arg Arg Arg Arg Val Cys Tyr Thr Val Leu Arg

Fig. 3.6 Sequences of amino acids in protamines from fish and bull sperm. (From Bradbury, E.M., *et al.*, *DNA Chromatin and Chromosomes*, Blackwells (1981), with permission).

tuna and clupeine in herring, but curiously carp and goldfish retain histone in sperm chromatin. Mammals such as human, horse, pig, sheep, ox and mouse have all been shown to have protamine in sperm, as has the chicken. The structure of nucleoprotamine is quite different to chromatin, and obviously gives greater protection to DNA and permits greater compaction. A quite elaborate series of steps are involved in the substitution of histone by protamine. This has been studied in the mouse, where it has been shown that variant forms of the normal histones H1, H2B and H3 are first synthesised and used to replace the normal somatic species. Intense histone acetylation follows, and then replacement by a sperm specific lysine rich histone. Finally, this is replaced by the arginine rich protamine in the native sperm. This process of gradual change and replacement, together with the conformational changes which accompany it, is termed spermiogenesis. Protamines are very small proteins of less than 50 amino acids, of which arginine makes up much more than half. Sequences of some protamines are shown in Fig. 3.6 from which it will be evident that the fish protamines are quite similar, but distinct from that found in bull sperm.

3.2 The Chromatin of Transcriptionally Active Chromosomes

Chromatin presents us with an unfortunate dilemma. Interphase chromatin, transcriptionally active as it is, is not readily visible by light microscopy and has proved remarkably resistant to resolution by electron microscopy. On the other hand, during mitosis and meiosis chromosomes are readily visible and amenable to study by both approaches, but sadly they are not transcriptionally active. Therefore the aspects of chromatin organization which relate particularly to its chief function, that of RNA synthesis, are in the main hidden from us. It follows that any special situations in Nature which afford us a look at transcriptionally active chromosomes must be especially prized. There are two such examples which will now be considered, the lampbrush chromosomes of vertebrate oocytes and the polytene chromosomes of some specialized tissues in larvae of Dipteran flies. More extensive discussion on both of these topics will be found in Bradbury *et al.* (1981) and Maclean *et al.* (1983); however, the former account, especially with regard to polytene chromosomes is somewhat dated.

3.2.1 The lampbrush chromosome

These structures are diplotene stage meiotic chromosomes and are found in the primary oocytes within the ovaries of a wide range of vertebrate animals. They have been most intensively studied in amphibians, especially newts. The primary oocyte has a very large nucleus, often termed the germinal vesicle, and its genetic content is 4C. Diplotene is very prolonged in the species involved, sometimes lasting for six months, and since the *Triturus* species of newts have a large genomic DNA content, the lampbrush chromosomes in these species are very big, often being as much as a millimetre in length. Chromosomes with a lampbrush-like conformation are not absolutely confined to vertebrate oocytes: somewhat similar structures are to be found in the unicellular green alga *Acetabularia*, and in certain chromosomes of insect spermatocytes. The name lampbrush alludes to the bottle brushes used for cleaning the interiors of lamp-glasses used on oil lamps, but, as a study of Fig. 3.7 will indicate, the name is not altogether apt, since the brush-like appendages in these chromosomes are actually loops rather than straight bristles. The loops vary greatly in size and character and are paired, since each chromosome is actually a doublet of chromatids, each of which exhibits an identical array of loops, (since the chormatids of a pair are genetically identical).

In order to understand the significance of the comparative loop morphology, it is necessary to remember that at meiotic diplotene, chromosomes exist as quadruplets, the paired homologous chromosomes being attached by the chiasmata. Although the loops of sister chromatids are identical, the loops of homologous chromatids sometimes differ, if there are particular differences at specific loci between the maternal and paternal homologues. The chromosomes pictured in Fig. 3.7 are one pair of bivalents only, from the newt *Triturus viridescens*, held together at four chiasma cross-over points.

When lampbrush chromosomes are studied in detail, it is found that each pair of loops emerge from a region of condensed chromatin, termed a chromomere. The relationship between loops and chromomeres is set out in Fig. 3.8, indicating that each chromatid is a single uninterrupted DNA duplex, (probably complexed with histone), and each loop simply an extension of this duplex, associated at each loop terminus by a region of more condensed chromatin. The chromatids therefore represent aggregates of chromatin that can be resolved into four separate regions, each forming one end of a single lampbrush loop. Chromosomes vary between 0.2 and 2 μm in diameter and there may be up to 1500 along one chromosome and up to 20 000 for the entire chromosome set. It is important to know how much DNA is in the loops, as a percentage of the total. Since each loop is approximately 50 μm long and the total loop number per haploid chromosome set averages 10 000, a total length of DNA in all loops combined would be 50 cm. But since such an amphibian would have not less than 500 cm of DNA in its haploid genome, it can be seen that about nine tenths of the total DNA is packed into the chromomeres. Actually it is the corollary of this figure which is the greatest surprise. Since, as will be stressed in a moment, the loops represent regions of transcriptional activity, it follows

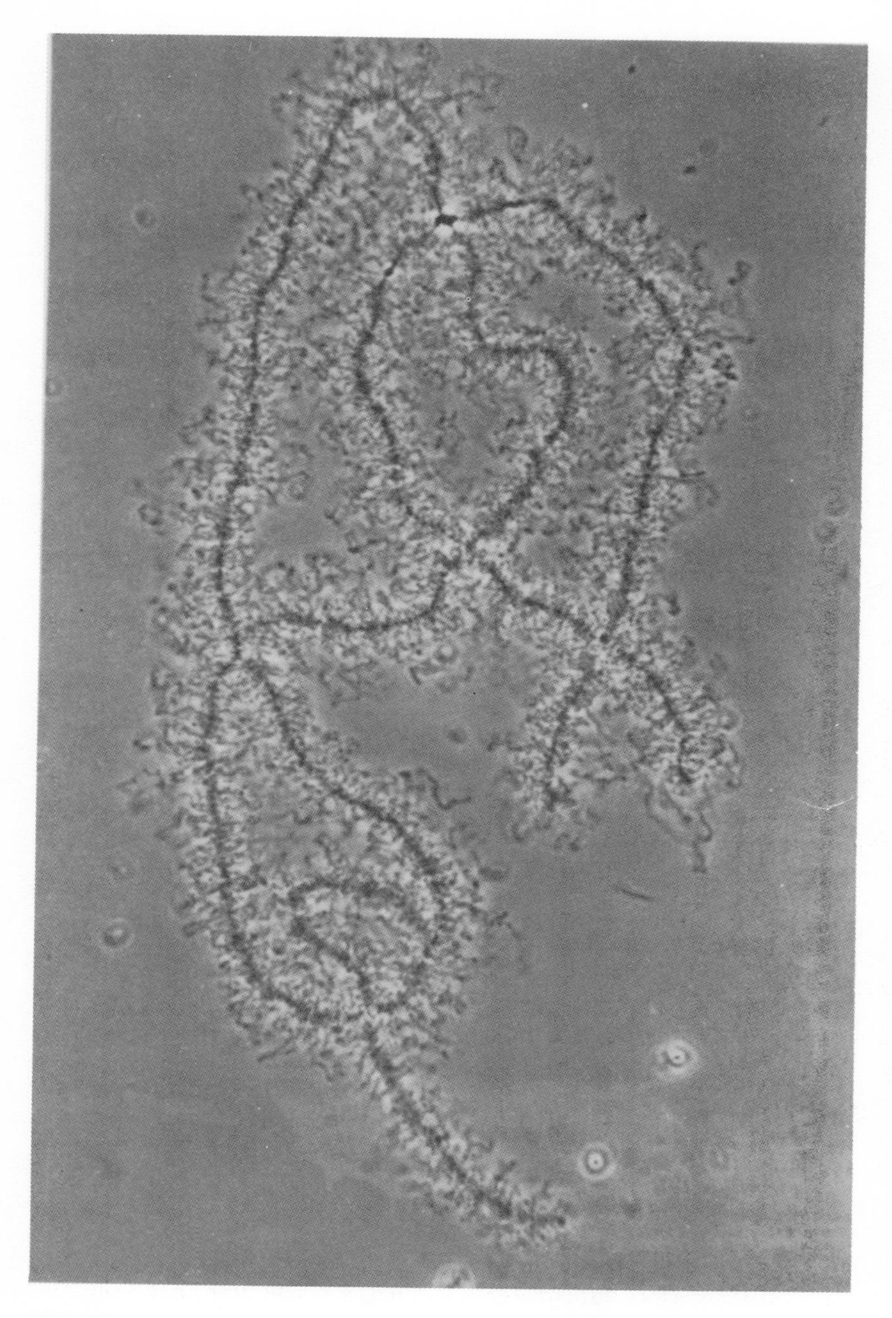

Fig. 3.7 A bivalent lampbrush chromosome from the newt *Triturus viridescens*. Notice the numerous loops of nucleoprotein extending from the central axis. Phase contrast microscopy at X500. (Courtesy of Professor J.G. Gall).

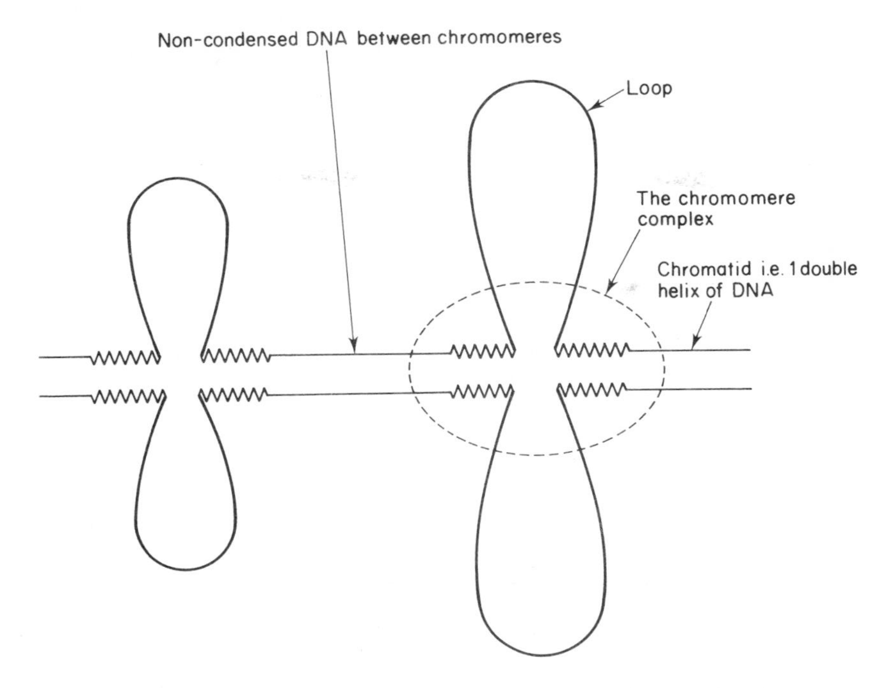

Fig. 3.8 Probable organization of DNA in a lampbrush chromosome, showing the extended DNA of the loops and the linking axis and the highly condensed chromomeric DNA. (From Bradbury, E.M., *et al.*, *DNA, Chromatin and Chromosomes*, Blackwells (1981), with permission).

that as much as one tenth of the genome is being transcribed at this stage of development, a surprisingly large percentage.

A closer look at the loops

In terms of the morphology of the chromosome, we have deduced that the DNA is highly compacted for the most part in beads of chromatin called chromomeres, and each of these regions of compaction is associated with a localised loop of decondensed chromatin. Unfortunately it remains unclear whether the chromatin in both chromomeres and loops retains its nucleosomal organization – the loops tend to be so heavily shrouded with the RNA products of transcription that any persisting nucleosomes are obscured. But on the other aspect of loop morphology, the transcriptional pattern, the picture is clearer. It is evident that most loops are in the form of 'Xmas trees', a name given originally to the appearance of the transcriptionally active large ribosomal genes of amphibian oocyte nucleoli. Each loop Xmas tree has a thin and thick end, the former being the region of initiation, the latter the region of termination, and the 'branches' being the primary product of transcription, in eukaryotes, the

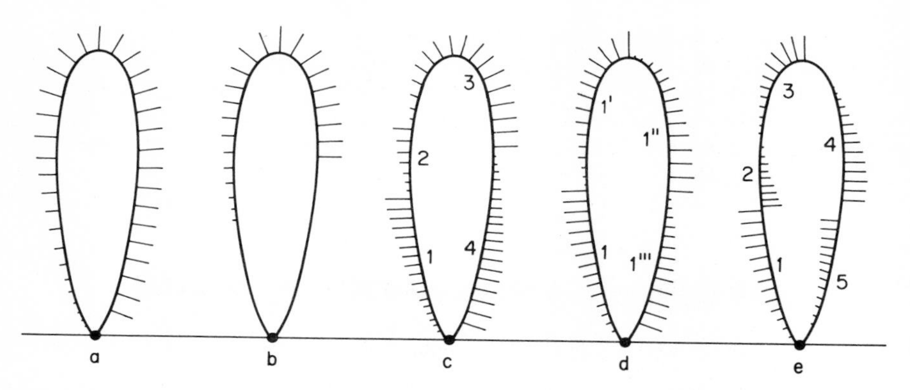

Fig. 3.9 Alternative patterns of transcriptional units in lampbrush chromosome loops. These patterns are deduced from phase contrast observation on actual loop morphology. Units numbered 1 to 1''' are presumed to be of equal length and perhaps to be repetitious copies of the same or similar sequences, while units 1 to 5 are of different lengths. Those in e show transcription from both strands but non-overlapping, so that 1, 3 and 4 are RNA products resulting from transcription of one strand in one direction, and 2 and 5 are RNA products of transcription from the other strand and in the opposite direction. (Reproduced by kind permission of Dr. V. Scheer).

hnRNA (heterogeneous nuclear RNA) which is the precursor of the cytoplasmic mRNA (messenger RNA). However, this is something of a generalisation, since, as indicated in Fig. 3.9 some loops have variable patterns, indicating that they contain more than one gene and, in some cases, that genes on one loop may be on different DNA strands and therefore are read in opposite directions.

Some conclusions about the lampbrush chromosome

a) The first conclusion being, if the lampbrush is taken as a model of chromatin organisation in transcriptionally active interphase chromatin, that active chromatin is decondensed and inactive chromatin condensed, at least as a general rule.

b) Not all genes are read from the same strand, and as expected, genes read from opposite strands do not overlap.

c) Active genes in these chromosomes are in very extended regions of DNA, and here nucleosomes may be absent, or in some cases in an altered conformation.

In reality the problems left by studies on lampbrush chromosomes are if anything more substantial than the light they shed on molecular aspects of gene organisation and expression. It is not clear why such a large part of the genome is transcriptionally active in the oocyte, nor what classes of sequences are represented by the condensed material in the chromosomes. Both repetitious and unique sequence DNA is known to be included in some loops, including sequences not transcribed in more representative cell types. Some puzzles are also afforded by comparing the lampbrush chromosomes of different amphibian species with different genome sizes, such as *Xenopus laevis* and the newt *Triturus carnifex*, the latter species have nine times the genome size of the former. As seen in Fig. 3.10 not only is the chromosome length much larger in the

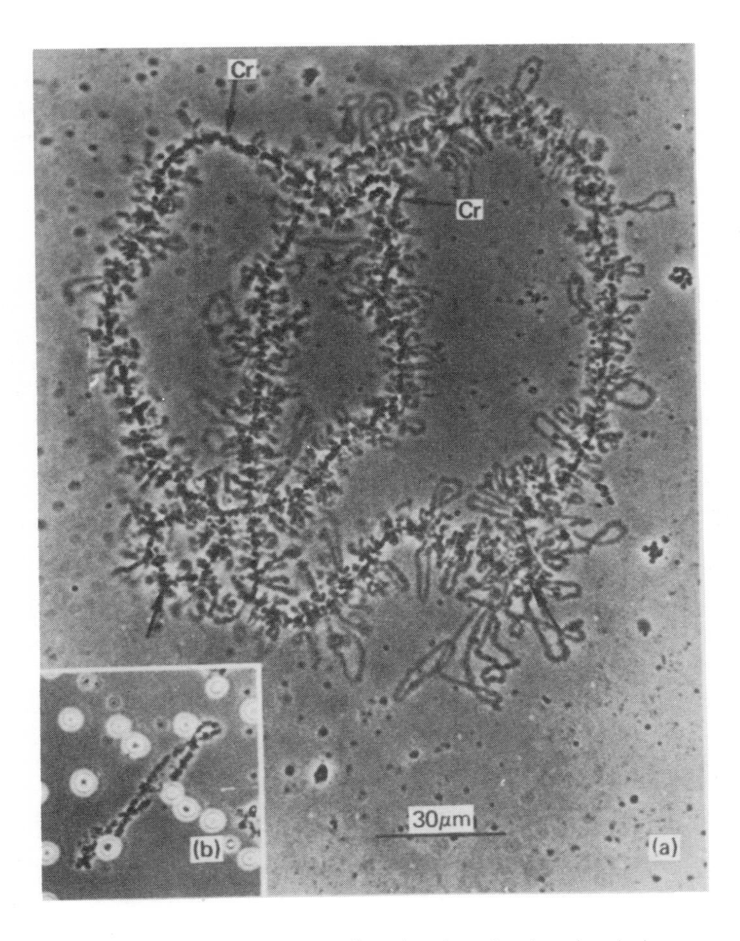

Fig. 3.10 Phase contrast micrograph of two bivalents in a lampbrush chromosome set. Those in (a) are from *Triturus carnifex* and in (b) from *Xenopus laevis* and both are at the same magnification. This illustrates graphically the way in which these chromosomes are the result of enlarged genome size, since the former species has a C value of 27 pg of DNA and the latter of 3. The areas in (a) marked Cr are centromeric bars which lack lateral loops. (From Vlad, M.T., Morphology of amphibian lampbrush chromosomes, in N. Maclean *et al.* (eds), *Eukoryotic Genes*, Butterworths (1983), with permission).

latter, but so is the average loop length. In summary, a look at the lampbrush chromosome leaves us with more problems than it solves.

Almost with relief we now turn to the second example of a transcriptionally active chromosome.

3.2.2 The polytene chromosome

As compared with the lampbrush chromosome, which is as we have seen, a transcriptionally active meiotic chromosome, polytene chromosomes are representative of true interphase chromatin; the only reason they are visible as chromosomes is that many rounds of DNA replication, (up to thirteen times the normal chromosome set in some species, and all undergone without any nuclear division) produces a ribbon-like structure which, although no longer than the extended DNA of a normal chromosome in the same species, is enormously broad. In addition, the replicated copies of DNA lie in parallel register in the cell, allowing determination of the distribution of DNA sequence types along the chromosome at a relatively gross level.

Giant polytene chromosomes are best known in the salivary glands, rectum, mid gut and some other tissues of larvae of two winged flies (Diptera), such as *Drosophila*, *Chironomus*, *Sciara* and others. Somewhat similar structures are found in a few other eukaryotic cells, for example, in the nucleus of the protozoan cell *Euplotes*, but from the point of view of modelling eukaryotic chromatin in action, those of *Drosophila* cannot be improved on. The cells in the larva in which these chromosomes are found are no longer capable of division, but the chromatin is none the less in the interphase stage of the cell cycle, and is transcriptionally active. As indicated by Fig. 3.11 the chromosomes present a strikingly banded appearance when viewed by phase contrast (or by bright-field microscopy after staining). However, before we discuss the significance of this banding, some aspects of the general chromosome morphology should be cleared up. Firstly, why are all the chromosomes linked together? The explanation is that in *Drosophila* there is a chromocentre to which all the centromeric regions are linked, leaving the arms free. There are only four chromosomes, chromosome 1 (the X chromosome), chromosome 2, with long left and right arms, chromosome 3, also with long left and right arms, and a small scarcely discernable chromosome 4. But where are the homologous pairs of chromosomes expected in a diploid cell? Again these chromosomes are aberrant in that both homologues lie in parallel and in register, so are not readily distinguishable. This has provided cytogeneticists with a delightful gift, because any deletions, inversions or translocations involving (as most do) only one homologue, will introduce a visible loop or irregularity in the other homologue when both homologues associate in this way. So many such chromosomal mutations have been intensively studied in polytene chromosomes. Thus in the bottom right of the picture in Fig. 3.11 a small portion of the chromosome arm 2R, near to its extremity, reveals a section where the two homologous chromosome copies have slightly separated.

Having sorted out the gross morphology, let us return to consider the bands

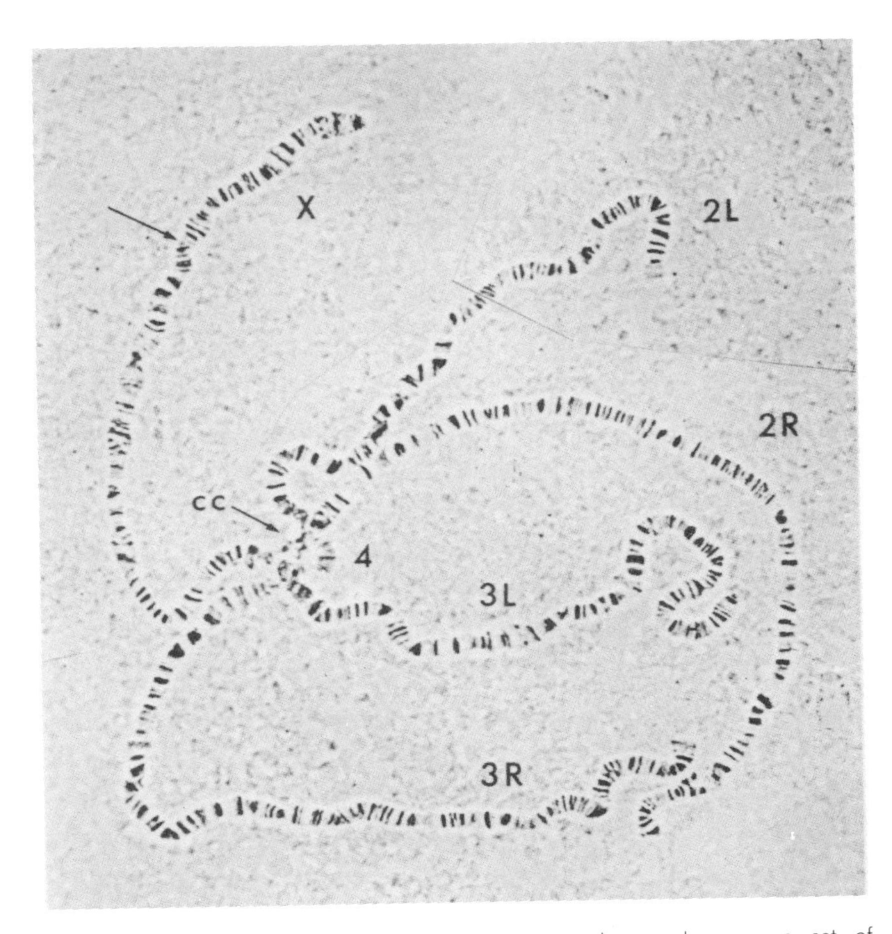

Fig. 3.11 A photomontage showing the entire polytene chromosome set of *Drosophila*. 2L and 2R are the left and right portions of chromosome 2, 3L and 3R are the left and right portions of chromosome 3, 4 is the short chromosome 4, and X marks the single arm of the X chromosome. CC marks the chromocentre. Notice the pattern of bands and interbands resulting from the numerous DNA strands in the chromosome ribbon being in precise register. (From C.B. Bridges and Ashburner, M., *The Genetics and Biology of Drosophila*, Academic Press (1978), with permission).

of these remarkable chromosomes. It should be explained that these bands are really chromomeres, that is they represent areas of condensed chromatin. They are therefore analagous to the chromomeres which subtend the loops of lampbrush chromosomes: also to the beads of chromatin, styled chromomeres, which appear along the length of leptotene stage meiotic chromosomes. These chromomeric bands must therefore not be confused with the bands which can be produced in mitotic chromosomes by Giemsa or Quinacrin staining after acid or alkali extraction. Such bands will be discussed briefly later in the context of the mitotic chromosome.

Polytene chromosome bands provide a chromosomal pattern which is essentially the same for chromosomes taken from different tissues in the same larva. So they do not represent, in themselves, a pattern which depends on selective gene expression. A glance at the pattern suggests that the amount of material in the bands is roughly the same as the amount in the interbands, but this is a very misleading view. It is misleading because a much greater degree of compaction is involved in the chromomeric (band) chromatin than the interchromomeric (interband) chromatin. Some calculations by Laird (1980) suggest that about 75% of the DNA of these chromosomes is in the bands and only 25% in the interbands. The total number of bands is similar to the chromomere number in the *Triturus* lampbrush, about 5000 *in toto* for the Drosophila chromosome set.

The most crucial question to ask of the polytene chromosome is where are the active genes? One way to find an answer is to provide cells with tritiated uridine as a precursor of RNA, and then to make autoradiographs to ascertain the locations of RNA synthetic activity. This experiment gives some interesting results, the first being that intense RNA synthesis is correlated with puffing, the dramatic decondensation of individual bands. Puffing has long been believed to be indicative of gene transcriptional activity, especially as the pattern in which bands are puffed varies from tissue to tissue. However, not more than 50 puffs can be detected at any one time over the entire chromosome set. We could therefore conclude that puffs represent the relatively dramatic activity of cell-specific genes, that is genes coding for products, such as globin, fibroin, or crystallin, that are made only in specialised cells but in large amounts (although none of these actual proteins will be made on the Drosophila cells involved). But if puffs represent the activity of cell-specific genes, where are the housekeeping genes whose products are needed in moderate amounts by all cells? It seems certain that they are not located in the bands since these areas are most often highly condensed.

If we return to the autoradiography experiment, but ignore the dramatic incorporation of label over puffed bands, then it will become evident that there is low level incorporation over all the interbands. It has also been found that antibodies raised against RNA Polymerase II (which transcribes messages), tagged with a fluorescent dye to render them detectable, will bind to these interband regions, again suggesting activity in transcription. Perhaps these uncondensed chromatin regions, which form an invariant pattern from tissue to tissue, are really the locations of genes that are constitutively active in all cells at all times. Such is the conclusion of those who have most recently and critically examined the evidence (Bautz, 1983 – see chapter in Maclean *et al.*,1983).

Let us now summarise what seems to be happening in the polytene chromosome. Bands account for three quarters of the DNA and in most cases are highly condensed. When a band is puffed the whole chromosome region seems to be involved in both massive decondensation and substantial transcriptional activity, and both are essentially tissue specific. Particular bands can be persuaded to puff experimentally by stimuli such as heat shock or steroid hormone treatment. Therefore bands are the locations of tissue specific genes. But a puzzle remains concerning puffing, since the DNA of a band is long enough to code for say twenty genes, yet a puff usually involves decondensation of an entire band. The reason for such massive chromatin decondensation remains obscure.

The constant open conformation and low level transcriptional activity of the so-called housekeeping genes is less surprising, although even here an interband is considerably longer than a gene. But as we will discuss in Chapter 4, perhaps both up and downstream flanking sequences need to be decondensed in order to ensure the satisfactory transcription of a structural gene. Before leaving the polytene chromosome, one further observation deserves mention, namely, that puffing appears to be, at least in some cases, infectious. If the two homologous chromosome copies are more or less in register (as indeed they are) we might expect some genetic mutations on one homologue to appear as a situation in which a puff involved only half of the band, that is the half representing one homologue. Such a situation is not uncommon. But when it does appear, it has been observed that when the two homologus portions are slightly apart from one another, only one half is puffed, but when they are closely connected the puff involves both homologues. It is as if puffing resembles crystal formation, instigating a similar decondensation in neighbouring chromatin which, left to itself, would not puff. Perhaps this helps to explain why so great a length of chromatin becomes involved in a puff in the first place.

It is satisfactory to conclude that, even if the lampbrush chromosome seems to afford more puzzles than solutions, the polytene chromosome provides us with some positive ideas concerning the correlation between gene activity and chromatin conformation in eukaryotic nuclei.

3.3 The mitotic chromosome

Having gleaned what useful information we can from transcriptionally active chromosomes, it is appropriate to look at mitotic chromosomes and learn what we can from this situation of transcriptional inactivity. The most important single observation in the present context is the most obvious one, namely, that mitotic chromosomes do not continue to synthesise RNA, and therefore there is a strict correlation between extreme chromatin condensation and transcriptional inactivity. It is as if the condensation of the polytene chromosomes chromomeres were extended to encompass all of the chromatin. So mitotic chromosomes are neat bundles of chromatin which are tied up securely so as to facilitate the chromosome sorting mechanism achieved by the mitotic spindle. But even in these highly compacted structures, as shown in Fig. 3.5, the DNA is distributed within the chromosome as loops extending from a scaffold, the latter extending along the axis of the chromosome. Presumably this scaffold or loop arangement persists even in the presence of nucleosomes and when the chromatin is at least partially dispersed during cellular interphase.

While on the topic of mitotic chromosomes, a short comment on mitotic chromatin banding seems appropriate. As illustrated in Fig. 3.12 when mitotic chromosomes are treated with agents such as trypsin, acid or alkali which are

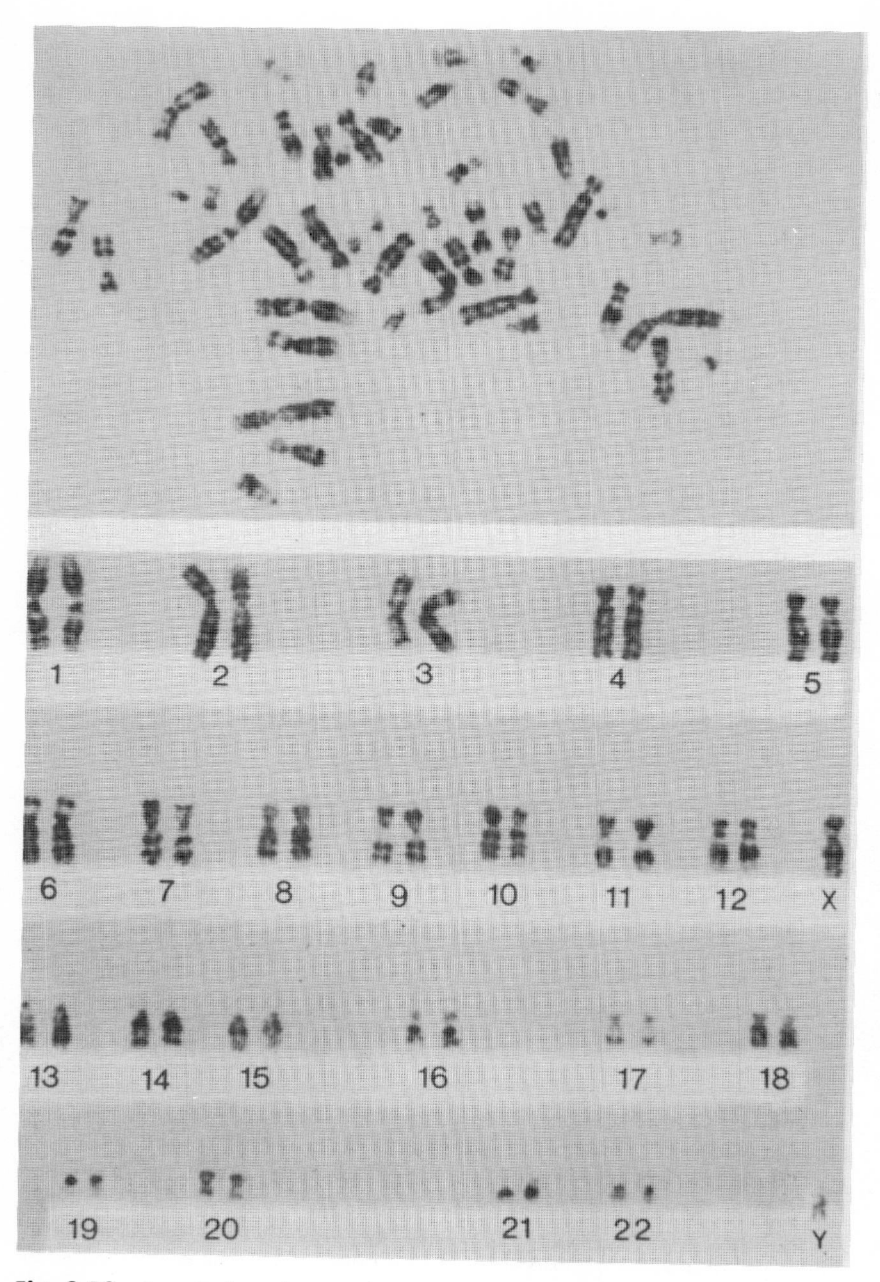

Fig. 3.12 A metaphase human chromosome spread and resulting idiogram after chromosome images had been cut out and arranged in order. All chromosome pairs have unique banding patterns due to the Giemsa staining technique used, while pairs of homologous chromosomes have identical banding patterns. (From Evans, H.J., *British Medical Bulletin*, **29**, 196 (1973), with permission).

likely to interact with histone, and then subsequently stained, a pattern of gross banding appears. At most there are only 20 bands per chromosome, and often far fewer, so it is banding on a quite different scale to that of chromomere distribution. The molecular basis for this differential banding pattern remains unclear, but all that needs concern us is that the technique is widely used for chromosome identification in routine cytogenetics. Indeed the technique has transformed human cytogenetics since it permits unequivocal identification and categorisation of chromosomes and the definitive localisation of translocations and deletions within the genome.

3.4 Interphase chromatin

Until now I have referred to interphase chromatin in normal eukaryotic cells as if, for the purposes of our present enquiry, it were a lost cause. But this is not entirely fair, and it is only proper that, before concluding this chapter a careful assessment is made of the organisation of interphase chromatin. After all, even if it is not amenable to microscopy, other techniques are available to permit useful enquiry to be undertaken.

3.4.1 Heterochromatin and euchromatin

This terminology was originally introduced to distinguish between chromosomes such as the Y and inactive X chromosomes of the human which remained visible and stainable during interphase as a result of their extreme compaction (they were therefore called heterochromosomes, since they stained differentially from others, not so compacted) and other less compacted chromosomes. But the terminology required modification, since the Y chromosome was invariably heterochromatic while either copy of the X could become heterochromatic (the condensed X chromosome in the female is often called the Barr body, and is widely used for nuclear sex determination). The Y was therefore said to consist of constitutive heterochromatin, and the X of facultative heterochromatin. Heterochromatin was defined in comparative terms, by comparison with euchromatin, being more compacted, stainable during interphase, and genetically inactive. Unfortunately, the usage of these terms has gradually broadened so that relatively condensed chromatin of any sort is sometimes referred to as heterochromatin, but such use of the word is best avoided. Once more, however, as with mitotic chromosomes, extreme compaction is closely correlated with transcriptional inactivity. Of course it is essentially a chicken and egg situation. Is the Y chromosome condensed because it is transcriptional, inactive or vice versa. Since only one or two genes have been located on the human Y chromosome a causative role is not clearly evident. In the case of the human X chromosome, however, it is hard to resist the conclusion that chromosome condensation is used to achieve genetic silence. Small blocks of heterochromatin have been identified on particular parts of chromosome arms, and these small sub-chromatin fragments also help to emphasise that compaction can help prevent transcription. Although a small block of heterochromatin in the midst of a section of normal chromatin may contain no gene sequences (in this it partially resembles the human Y chromosome), it also has the capacity to impede the transcriptional activity of adjacent sequences, the so-called position

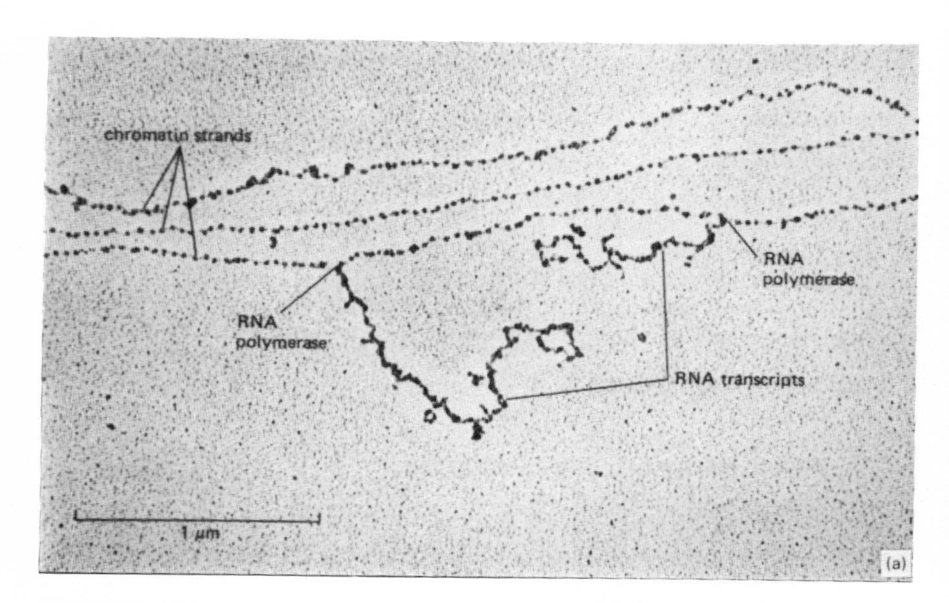

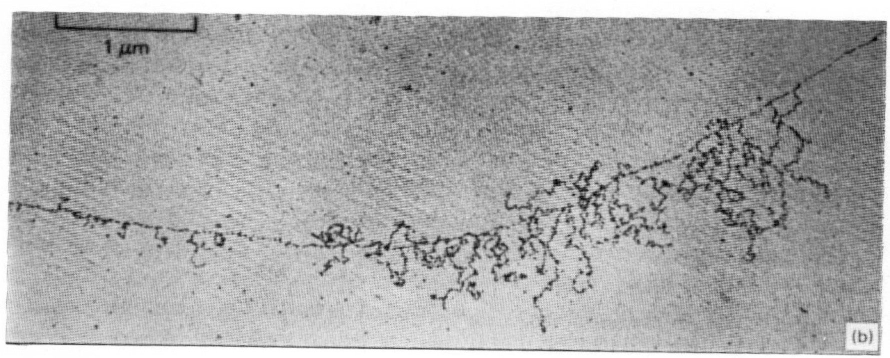

Fig. 3.13 (a) A series of three chromatin strands showing clear nucleosome arrays. In the lower of the three strands, two RNA polymerase II molecules are engaged in transcription and the partially condensed transcripts are visible. In the average eukaryotic cell only 1 molecule of RNA polymerase II occurs for every 750 nucleosomes (150 000 bases of DNA). (b) A region of chromatin in which RNA polymerase II molecules are transcribing a gene at high frequency and many growing hnRNA transcripts are visible. Such regions of actively transcribing chromatin are comparatively rare in the average eukaryotic cell nucleus, emphasising how selective gene expression actually is. ((a) Courtesy of Victoria Foe. (b) Reproduced, with permission, from Foe, V., *et al.*, *Cell*, **9**, 131 (1976)).

effect. Such small independent blocks of heterochromatin are termed intercalary heterochromatin.

3.4.2 Active genes and the nuclear lamina

Applied to the inner surface of the nuclear envelope is a thin laminar sheet consisting of a fibrous protein network. This lamina (also called the nuclear cage) seems to fulfil a number of roles, one being the organisation of the envelope itself, and the other the structural support of the nuclear pore complexes. But for us its most interesting role is that it is frequently associated with particular parts of interphase chromosomes. Although chromosomes are not readily visible during interphase (with certain exceptions already referred to), there is evidence that interphase chromatin retains its approximate three-dimensional chromosomal form both relative to the nucleus as a whole and relative to other chromosomes. It also seems that individual chromosomes are not randomly located in the nucleus, but retain a position relative not only to other chromosomes, but more particularly relative to the nuclear lamina. However, the latter relationship is dynamic, and it is quite possible that certain portions of chromosomes enriched in active genes are brought by the interplay of the nuclear lamina into the proximity of nuclear pore complexes, at least for as long as these genes are to remain active. This laminar cage is also thought to have an important role in ordering the positioning of chromatin during DNA replication.

3.4.3 Interphase chromatin and nucleosomes

Electron microscopy of spread films of chromatin has enabled visualisation of transcriptionally active chromatin to be achieved in some cases. What can be learned from these observations? The first conclusion is that transcription is a rare event in comparative terms within the bulk chromatin of an interphase eukaryotic cell. In chromatin spreads, most of the DNA is not apparently active in RNA synthesis. Of the genes that are clearly being transcribed, many are being transcribed at a relatively slow rate (see Fig. 3.13). Such genes are in an extended chromatin conformation but are still organised around nucleosomes. Some genes, such as those for ribosomal or transfer RNA, or those coding for products required in great quantity by the cell, are clearly quite unrepresentative of bulk chromatin, but they provide us with the remarkable but now familiar 'Xmas tree' pattern of closely packed precursor RNA (see Fig. 4.2). In a few of these situations it seems very questionable whether nucleosomes persist during such mass transcription. The two situations are presumably analagous to interband transcription and puffed band transcription in the insect polytene chromosome discussed in Section 3.2.2.

4

Mechanisms of Gene Regulation

4.1 The Need for Regulation

Before examining the mechanisms used by cells and organisms to regulate their genes, it is appropriate to think first about the need for such regulation. It is conceivable that in an organism with a small genome, all the genes could be transcribed continuously at maximal rate without any biochemical inconvenience. But in all situations, except perhaps the very simplest viruses, gene regulation is an essential aspect of success. We saw that in the case of phage lambda (discussed on page 35) there is a choice between lysis and lysogeny, and this choice is made by controlling the synthesis of particular products by specific genes. Let us consider a bacterial species capable of sporulation. As seen in Fig 4.1 particular gene products are necessary for sporulation and it would clearly be a great biochemical and even a morphological problem to have molecules appropriate to spore formation produced in a cell that was at a stage of growth inappropriate to sporulation. When the enormous genomes of many eukaryotes are considered the need for regulation is greater than ever, especially since cell specialisation, which is so characteristic an aspect of multicellular organisms, requires that cells differ in form and function from one tissue to another, yet in most cases, share a common gene pool that constitutes a diploid representation of the entire genome. The relationship between cell differentiation and gene regulation will be explored more fully in Chapter 5.

An important assumption has been made in the first paragraph, namely that the regulation of gene products in the cell, most of which in their functional and effective form are proteins, is a result of control at the level of transcription of DNA into RNA. It is therefore possible for gene expression to be regulated post-transcriptionally by mechanisms of selective messenger RNA breakdown, or at a translational level by selective message translation. There is no doubt that these latter levels of control are indeed important in individual cases. For example, it has been demonstrated that the build up of new histone in a cell to provide for the needs of DNA synthesised during the S phase of the cell cycle is a result, not of discontinuous transcription of the histone genes, but of variable breakdown rates of histone messenger RNA (Maxson *et al.* 1983). This and other similar situations are probably exceptional. There is reasonable evidence

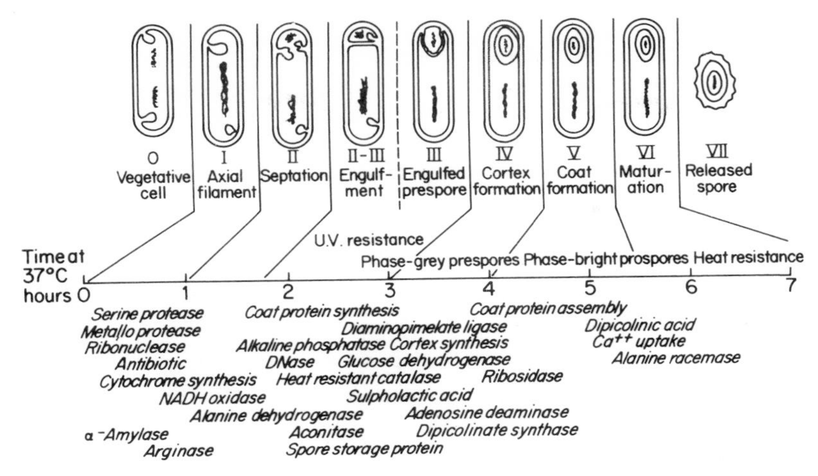

Fig. 4.1 The appropriate timing (horizontal axis) of the main morphological and biochemical changes which occur during sporulation in a *Bacillus* species. (Reproduced with permission from Mandelstam, J., *et al.*, *Biochemistry of Bacterial Growth*, third edition, Blackwells (1981)).

that, at least in most cases, the main mechanisms for regulating cellular gene expression are transcriptional, and indeed the polytene chromosomes of *Drosophila*, discussed in Chapter 3, provide visual evidence for this in that puffing patterns, unlike chromomere patterns, are tissue specific and variable with times of development in any one tissue. Moreover, since this book concerns genes it is also best to confine the discussion to those aspects of control nearest to the gene, that is the transcriptional level.

4.2 How a gene works

Since understanding gene regulation requires an appreciation of the basic mechanics of transcription, it is appropriate to discuss how transcription proceeds in prokaryotic and eukaryotic cells, even in the absence of specific regulation. As established in Chapter 1, in all cellular organisms a gene consists of a single strand of DNA (the sense strand) hydrogen bonded to form a DNA duplex to another complementary strand (the antisense strand). Transcription requires the synthesis of a growing strand of RNA in complementary sequence to the sense strand DNA, and, since the DNA is polar, the RNA can only by synthesised in a 5′ to 3′ direction, implying that the (sense) strand of DNA is read in a 3′ to 5′ direction. The enzymes chiefly responsible for catalyzing this synthesis are called RNA polymerases. These are large multi-subunit proteins. Only a single species of this enzyme exists in bacterial cells, but three distinct enzymes are found in the nuclei of eukaryotic cells and a fourth type occurs in mitochondria. Since *E. coli* has approximately 6000 molecules of RNA polymerase, and eukaryotic cells, on average, about 40 000 molecules of RNA polymerase II (which is responsible for production of mRNA) it is clear that the availability of this enzyme is itself a very important regulatory mechanism in gene expression. The three main eukaryotic enzymes are styled RNA polymerase I, II and III, and these are responsible for production of large ribosomal RNA, messenger and message precursor RNA, and transfer RNA and small ribosomal RNA respectively.

RNA synthesis consists of four discrete stages. These are

1. binding of RNA polymerase to a specific recognition site
2. initiation
3. RNA chain elongation and
4. chain termination and release.

Let us first consider the situation in bacteria, where only one species of RNA polymerase exists. How does the enzyme find the end of a gene at which it may commence transcription? The answer is that upstream (that is 3′ to the start of the sequence of the sense strand) from all genes are to be found short sections of DNA known as promoter regions. Entire promoter regions may be up to 200 bases long and are rather variable between different genes. But all bacterial promoters contain a so-called consensus region of about seven base pairs known as the Pribnow box. The sequence is TATAATG, or something very close to it, and lies between six and nine bases above the transcription start point of the gene. Another bacterial consensus region within the promoter lies even further upstream and is rather more variable, but clearly the enzyme complex RNA polymerase homes in on these sequences and is then guided in some way into the initiation point itself. The later steps of RNA chain elongation and termination are scarcely relevant here and are discussed in all textbooks of biochemistry. It will be seen that in the absence of other factors, all bacterial genes are essentially in open competition for a somewhat limited pool of RNA polymerase enzymes. Depending on which gene is transcribed, the same bacterial enzyme is responsible for production of ribosomal, transfer or messenger RNA, the latter species then engaging with ribosomes for the final process of message translation into the amino acid chain of a protein molecule.

Turning now to eukaryotes, the process of transcription is essentially similar but there are a number of additional complications. These can be itemised as follows:

1. Three different classes of RNA polymerase are present.
2. As compared to bacterial cells in which mRNA turns over rapidly, most eukaryotic mRNA is long lived, a single molecule having a life of many hours, days, or even weeks.
3. Many eukaryotic genes contain introns. The initial product of transcription is therefore a large precursor RNA; in the case of messenger RNA this is the

heterogeneous nuclear RNA (HnRNA). Post-transcriptional processing leads to excision of intron sequences in the precursor RNA and ligation of exon sequences together, to yield the final product.

4. Eukaryotic RNA is modified after transcription by capping and tailing. The former consists of the addition of a methylated G residue to the 5′ end, and the latter the attachment of a string of up to 200 adenylic acid residues to the 3′ end.

5. Many bacterial genes are transcribed in tandem into a large polycistronic messenger RNA. This is not normally a feature of eukaryotic transcription.

6. As with bacterial genes, there are promoter sequences recognised by eukaryotic RNA polymerase enzymes. Polymerase I recognises a series of short sequences which lie in the so called spacers between the multiple tandem copies of the large ribosomal genes but there is only one major promoter per gene. In the case of polymerase II, the situation is basically similar to the bacterial one, there being CAT and TATAA boxes, functioning as attractant recognition sites for enzyme. These consensus sequences are usually located about 50 and 25 bases upstream from the initiation start site of the gene. Remarkably the polymerase III enzyme recognises a regulatory protein that binds to a promoter in the middle of the coding sequence itself. Clearly the large polymearse III enzyme has to 'reach over' to recognise both the promoter and initiation point together. With all the eukaryotic polymerases, there are proteins that act as transcription factors and help the polymerase enzyme recognise the promoter.

7. Most eukaryotic organisms are multicellular and display considerable cell specialisation as a concomitant of this multicellularity. As emphasised in Chapter 5, such cell specialisation involves selective gene expression, so that in any one differentiated cell, only a limited portion of the entire genetic potential is ever expressed. If we look in detail at the kinds of gene sequence which are expressed in different cell types, we will find that, as a generalisation, there are two distinct patterns. Some genes are expressed in all cells. These are genes that code for the so-called housekeeping proteins, that is molecules that, as a result of their fundamental role in cell structure or metabolism, are essential to the survival of all cells. Other genes, coding for cell-type specific proteins, are differentially expressed, and transcripts for such genes can only be found in cells within which that particular product has a selectively relevant role. So both cytoplasmic mRNA and nuclear hnRNA will vary in both content and complexity from itssue to tissue in accordance with the orchestration of genes working for both housekeeping and cell-type specific proteins.

Having outlined the essentials of the process of transcription in eukaryotes, let us go back and detail the characteristics of some standard examples of eukaryotic genes within the genome, remembering that all of them are complexed with nucleosomes and therefore, even during transcription, are most probably in the form of chromatin. Here, however, we will concentrate on the

DNA. The two genes which will be examined in detail are the large ribosomal RNA genes of *Xenopus laevis* and the beta globin gene of the mammal. Both of these sequences have been intensively studied and have recently been reviewed by Moss, Mitchelson and De Winter (1985) and Flavell and Grosveld (1983) respectively.

4.3 Structure of large ribosomal RNA genes of Xenopus laevis

Xenopus laevis has 450 copies of the transcription unit for the large ribosomal RNA, all of these copies being arranged in a single tandemly arranged series. The DNA carrying these genes is known as the nucleolus organizing region – the nucleolus constitutes the amassed products of these gene sequences before they are exported to the cytoplasm as contributary molecules to the formation of ribosomes. We need only concern ourselves with the structure of a single copy of this repeating unit of DNA. Following careful extraction and spreading, it is possible to visualise these sequences and their RNA products in the electron microscope, see Fig. 4.2. The whole transcription unit is

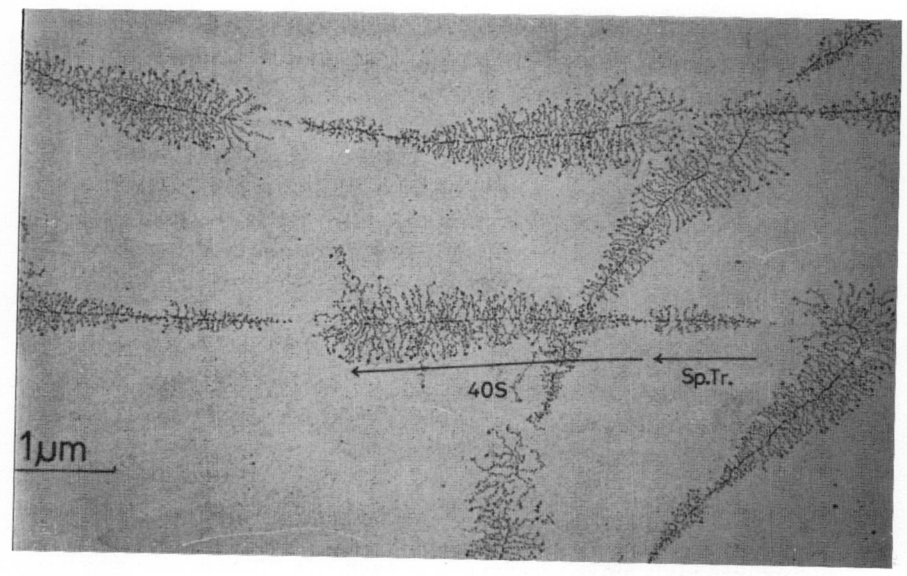

Fig. 4.2 Electron micrograph showing transcription of ribosomal genes in nucleolar material recovered from *Xenopus laevis* oocytes. 40S and Sp. Tr. refer to the 40S transcribed and spacer transcribed DNA regions. The 'Xmas trees' are RNA transcripts of the 40S ribosomal RNA precursor in various stages of completion. (From Moss, T., *et al.*, The promotion of ribosomal transcription in eukaryotes in *Oxford Surveys on Eukaryote Genes*, **2**, 207–250 (1985), with permission).

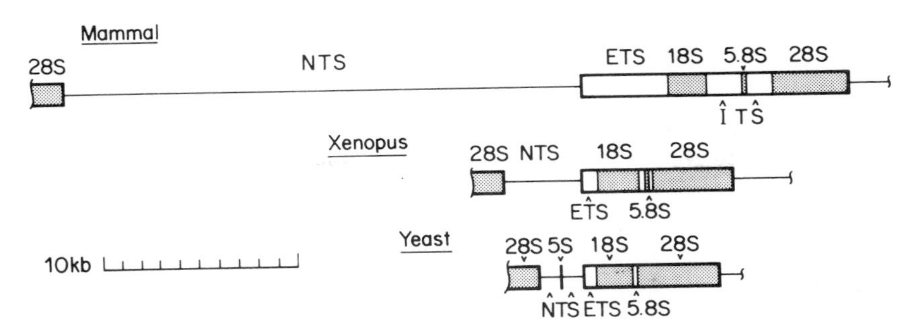

Fig. 4.3 Organization of the genes for ribosomal RNA in mammals, *Xenopus*, and yeast. Boxed regions are those transcribed, NTS is spacer that is chiefly not transcribed and does not contribute to the final product (except in *Xenopus*). ETS and ITS refers to external and internal transcribed spacers respectively. Transcription proceeds from left to right. (From Moss, T., *et al.*, The promotion of ribosomal transcription in eukaryotes in *Oxford Surveys on Eukaryote Genes*, **2**, 207–250 (1985), with permission).

transcribed by an RNA polymerase enzyme from end to end, and so the familiar Xmas tree pattern of RNA which can be seen to adorn the gene sequence consists of precursor ribosomal RNA transcripts of varying length, and these go to form the side branches of the tree. As a polymerase enzyme traverses further along the sequence, the RNA product becomes longer and so the 'branch' subtended by the enzyme is larger. It will also be noted in Fig. 4.2 that there are small RNA branches evident in the spacer regions between the main transcription units. These will be discussed below.

Figure 4.3 shows a diagram of the precise format and size for the sequences within a single transcription unit and spacer. The repeating unit is close to 12 kilobases long and each haploid chromosome set has, on one chromosome, 450 tandemly placed copies in a colinear DNA sequence. The three coding sequences are arranged in the order 18S, 5.8S, and 28S, these figures representing the sedimentation constant of the RNA species and therefore being roughly indicative of size. Although no intron sequences occur within the gene coding regions, there are so-called spacer sequences placed between the genes and, as shown in Fig. 4.4, the initial product of transcription is a 40S precursor RNA of almost eight kilobases. The point of initiation of the transcript is almost one kilobase upstream from the beginning of the coding sequence of the 18S gene. This so-called transcribed spacer, together with other spacer regions within the initial transcript, is cut out by processing enzymes to leave the final products shown in Fig. 4.4. Within the region immediately above the point of transcriptional initiation is a region of approximately 170 base pairs which is deemed to act as a promoter for the polymerase 1 enzyme, and is very similar in three different species of *Xenopus* (see Fig. 4.5) whose ribosomal genes have been cloned and extensively sequenced. In a region that is, for the most part, extensively made up of GC pairs, it will be seen that AT rich regions exist upstream and immediately around the transcription initiation site (+ 1 on Fig.

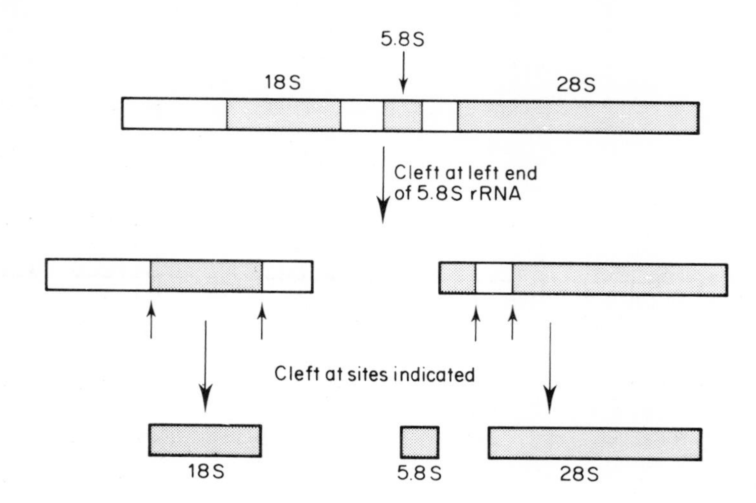

Fig. 4.4 Stages in the processing of the precursor molecule of the large ribosomal RNAs in mammals, with the final products shown in the lower part of the figure. (From D. Freifelder, *Molecular Biology*, Van Nostrand Reinhold, with permission).

4.5) and again between − 17 and − 30 bases upstream. As we shall see, this is analogous to the promoter region found upstream of initiation sites in all polymerase II transcribed gene (see section 4.4).

How do we know that the designated promoter sequence is indeed necessary for transcription, acting as a recognition site for the polymerase enzyme? A recently developed technique provides the answer. The methodology, known broadly as surrogate genetics, involves recovering the cloned sequence of the ribosomal transcription unit, together with flanking sequences, and inserting them into a plasmid vector. This DNA construct can be grown in bulk following further cloning, and then placed in an environment which provides all the necessary molecules for transcription. These requirements can be provided either by placing the sequences in a cell-free extract from *Xenopus* tissue culture cells, or by injecting the plasmid constructs into the nuclei of amphibian oocytes. A further elaboration on the experiment provides the vital information needed to answer our question. This is to manipulate the ribosomal RNA coding sequence by deletion of particular parts of the molecule and testing the remaining sequence for evidence of transcription. Using this approach it has been shown that deletion of the region between the initiation point and some 200 bases upstream does indeed eliminate transcription, helping to identify this region as a promoter sequence. Deletion of parts of the gene coding sequence, on the other hand, does not ablate effective transcription but simply reduces the length of the transcript.

Having identified the promoter sequence, the three coding regions and the transcribed spacers between the coding regions that are cut out during post-transcriptional processing, it remains for us to discuss the role of the long

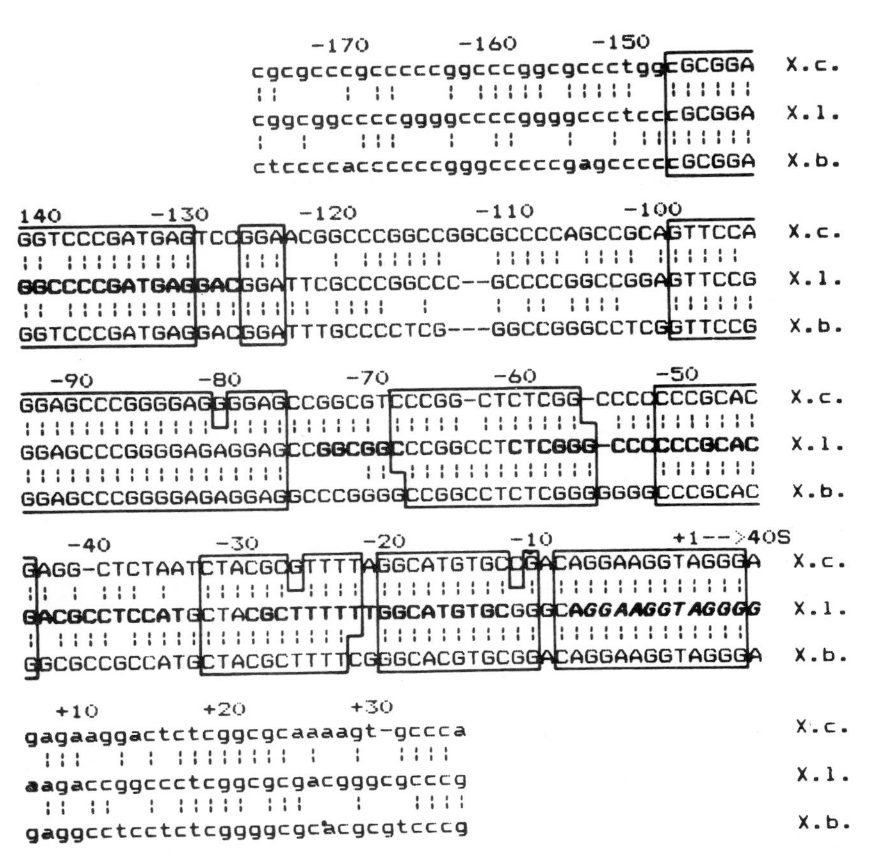

Fig. 4.5 Alignment of homologous promoter regions of the rRNA genes in three closely related amphibians *Xenopus clivii* (X.c.), *X. laevis* (X.l.) and *X. borealis* (X.b.). Regions of homology are boxed. 1 marks the start of the coding sequence. (From Moss, T., *et al.*, The promotion of ribosomal transcription in eukoryotes in *Oxford Surveys on Eukaryote Genes*, **2**, 207–250 (1985), with permission).

spacer which lies between the downstream end of the 28S coding sequence and the upstream end of the primary transcript. In *Xenopus* this is a region of between 3 and 4 kb, although in the mammalian genome it is very much longer. This region was originally referred to as the non-transcribed spacer region, but a close look at Fig. 4.2 will reveal that a small fringe of transcripts can be seen quite distinctly in association with the DNA of this region. The work of Moss *et al.* (1985), and of others whose work is mentioned in that review, indicates, again by surrogate genetics, that although most of this region apart from the promoter sequence lying within 200 bp of the initiation site of the primary transcript, is not essential for transcription. If present it greatly enhances the rate of transcription. Furthermore, it has been found that a series of promoter-like sequences are found within this extended spacer region, and these have been termed spacer promoters. They direct transcripts which terminate upstream of the main ribosomal promoter, and this can be verified by study of Fig. 4.2. A plausible hypothesis concerning the probable function of this spacer region is that it acts as a general attractant for RNA polymerase 1 enzymes, bringing them into the vicinity of the main ribosomal promoter just as the sight and scent of a flower attracts insects into the vicinity of stigma and stamens. The resulting transcripts from this region are currently thought to be redundant.

4.4 The mammalian beta globin gene

The ribosomal genes discussed in the section 4.3 are a prime example of a DNA sequence that is transcribed in all cells more or less constitutively. The globin genes, transcribed by RNA polymerase II, serve as examples of genes which are expressed in a strictly tissue-specific manner. They are a small family of genes which together exhibit all of the key aspects of eukaryotic genes coding for messenger RNA.

In mammals the genes coding for the alpha-like globin genes and beta-like globin genes are located on different chromosomes, but clearly their expression must be coordinated since, in the adult mammal, the alpha and beta globins are required in identical amounts.

The precise location of the beta and beta-like genes is best known for the human, and here, as shown in Fig. 4.6 a series of coding sequences are found on chromosome 11, all on the same DNA strand. These sequences code for epsilon globin, two variant forms of gamma globin, delta globin and beta globin respectively. Of these, epsilon is expressed only in embryonic life, both gamma globins during foetal life, and delta and beta during adult life (although the former only in small amounts). Each of these genes is approximately 1.5 kb in size, but, as will be stressed in a moment, most of this length is not actually DNA which is represented in the final messenger RNA. If we now consider the beta globin gene in detail, and take the rabbit gene as an example, its structure is shown in Fig. 4.7. The remarkable feature of this gene, and indeed most other eukaryotic genes, is that the DNA sequence expressed in the messenger RNA (the exon sequences) is split by two intervening sequences (the intron

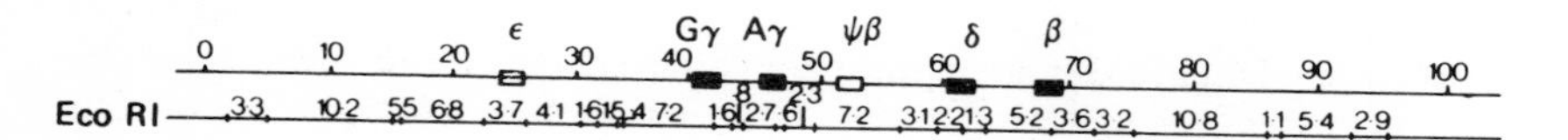

Fig. 4.6 General structure of the human beta globin region of the short arm of chromosome II. The upper line shows the gene position and order of, reading from left to right, the embryonic epsilon gene, the two copies of foetal gamma genes Y^G and Y^A, a beta like pseudogene, and the adult δ and β globin genes respectively. The lower line is the restriction map of the region using the enzyme Eco. RI. (Reproduced from Flavell, R.A. and Grosveld, F.G., Globin genes: their structure and expression, in N. Maclean *et al.* (eds), *Eukaryotic Genes*, Butterworths, (1983)).

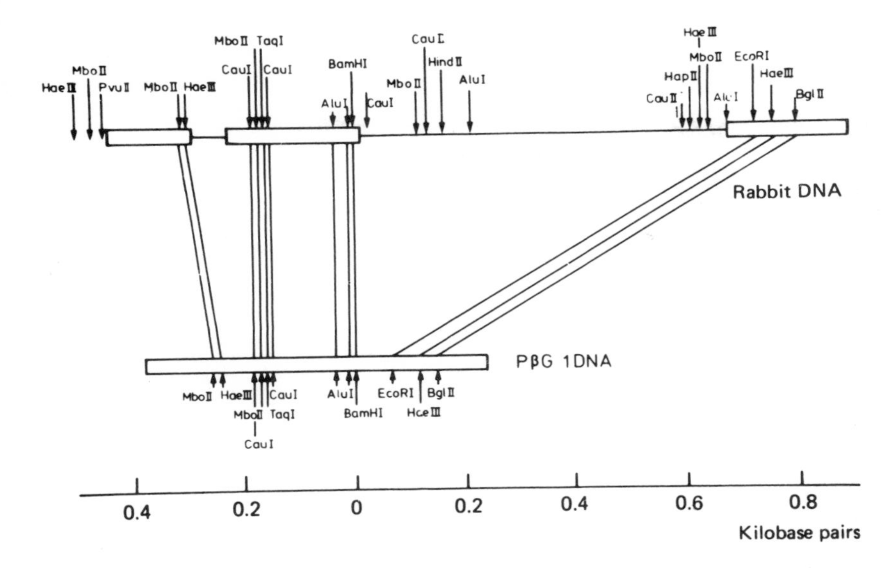

Fig. 4.7 Schematic drawing showing the structure of a typical globin gene, that for rabbit beta globin. Exons are shown as blocks, introns as connecting lines, and in the lower part of the figure is shown a complementary DNA copy derived by reverse transcription of the rabbit beta globin mRNA. (Reproduced from Flavell, R.A. and Grosveld, F.G., Globin genes: their structure and expression, in N. Maclean *et al.* (eds), *Eukaryote Genes*, Butterworths, (1983)).

sequences) which are not represented in the message. As will be clear from looking at Fig. 4.7, the intron sequences make up a greater length of the total DNA sequence than do the combined exons. In a situation analogous to that involved in the processing of the primary transcript of the large ribosomal RNA genes, the primary transcript of the beta globin gene is a high molecular weight molecule of the heterogenous nuclear RNA (hnRNA) class found so abundantly in the nuclei of eukaryotic cells. This large precursor is then cleaved within the nucleus and the exon sequences spliced together to form the beta globin messenger RNA molecule. The precise positions of introns correlate with discontinuities in the protein molecule itself, and these mark the boundaries of the distinct protein domains. Although intron splicing is carried out so precisely the actual intronic DNA seems to play no role in gene expression, and indeed surrogate genetic experiments demonstrate that the intronic DNA can be substantially deleted without affecting either RNA splicing or subsequent activity of the RNA message.

The DNA sequences at the 5′ and 3′ splice junctions show sequence conservation, revealing that there is a requirement for recognition, cleavage and ligation of the pre-mRNA molecule by the specific splicing enzymes. The general consensus sequences of these splice junctions are as follows. At the 5′ end

exon ← AG/GTAAGT → intron
exon ← CAG/N → intron
where N ± any base

The globin gene promoter

The RNA polymerase II enzyme is attracted to a sequence upstream from the initiation point in the structural genes of eukaryotes in a somewhat similar way to the function of the ribosomal gene promoter. However, there is only a single promoter region, although two consensus sequences within it are particularly conserved. These are the TATA box with a consensus sequence of TATAAAAG and the CAAT box, with a consensus sequence of GGCCAATCT, located at − 30 and − 75 bases upstream from the initiation point of the gene. Work with both regions of the promoter has been carried out with the rabbit beta globin gene. Loss of the CAAT box region simply decreased the rate of transcription of the gene, while deletion of the TATA box leads to inappropriate initiation of the polymerase at a range of sites instead of specifically at the proper initiation site. Thus the promoter region clearly functions to attract and bind RNA polymerase molecules and to guide them so that the RNA transcript molecule is started at precisely the right place.

In addition to the beta globin gene having a transcriptional promoter region, recent evidence indicates that other regions far upstream also affect the efficiency of transcription, especially in relation to particular tissues, so apparently helping to ensure the tissue specific expression of this gene. Such regions have now been termed *enhancer sequences* and, although at first found only in association with viral genes, there is increasing evidence for their presence and importance in relation to the expression of many different eukaryotic gene sequences (see review by Picard, 1985).

Globin Pseudogenes

While considering the molecular biology of the mammalian globin genes, it is useful to study another aspect of interest that occurs in association with other families of genes within eukaryotic genomes. Pseudogenes were discovered only by hybridizing short molecules of DNA coding regions with partially degraded genomic DNA. It became clear that some regions hybridized but were not represented in the population of known protein products. If the mammalian globin genes are an accurate indicator, then there are as many pseudogenes in eukaryotic genomes as there are normal genes. Pseudogenes are characterised by two or three particular qualities. The first is that they frequently contain stop codons within the coding sequence, which would terminate translation early. Secondly, they most frequently have no introns, even when the genes to which they are so closely related possess introns. Intronless pseudogenes are said to be processed, and are presumed to have arisen by DNA synthesis from a messenger RNA molecule (which of course also lacks introns). At present the only enzymes known to be capable of synthesising DNA from an RNA template are the reverse transcriptase enzymes of retroviruses. Perhaps a long evolutionary relationship between the cells of eukaryotes and infecting retroviruses accounts for these surprising genetic elements. (See also discussion of the pseudogenes on page 3 of Chapter 1).

4.5 Mechanisms of Gene Regulation in Prokaryotes and Eukaryotes

Having discussed the structure and the mechanics of expression of two different examples of eukaryotic genes, we are now better equipped to consider the mechanisms which are employed to regulate such genes. While the ribosomal RNA genes may be expressed more or less continuously and their rate of synthesis dependent chiefly on the supply of polymerase I type enzymes, it is clear that for genes such as beta globin, the situation must be quite different. Here is a gene which is inactive in all cells during embryonic and foetal life, and inactive in most tissues in adult life. But in one tissue, the erythroid cells of the blood, this gene is highly active. Obviously mechanisms quite different from the general availability of RNA polymerase II are determining its activity or inactivity in particular cells. And although bacteria are not multi-cellular and therefore do not display tissue specific gene expression, they nevertheless have mechanisms to ensure that not all genes are expressed at all times. We will begin with a look at gene regulation in bacteria and then progress to the more complex eukaryotic situation.

4.5.1 Bacterial operons

In the late 1950s two French scientists, Drs. Jacob and Monod, were investigating a remarkable property of *E. coli*, the phenomenon of induced enzyme synthesis. It became evident that synthesis of an enzyme necessary for the metabolism of a substrate could be cut off by the bacteria in the absence of the substrate. Moreover, in some situations where a series of enzymes were all involved in different steps of one metabolic pathway, absence of the primary substrate could lead to the shutting off of synthesis of all the enzymes at once. The corollary also applied, in that when the substrate (or some other trigger factor active as an inducer) became available, all enzymes were synthesised together. This is referred to as *coordinate regulation* and the whole process as *induced enzyme synthesis*. What Jacob and Monod discovered was that such coordinate regulation and induction resulted from the structure and operation of a particular genetic complex, the operon, and their studies on the lac operon of *E. coli* were published in 1961 (Jacob and Monod (1961)) and resulted in their jointly receiving a Nobel prize some years later.

The essentials of the operation of the lac operon are that it consists of a series

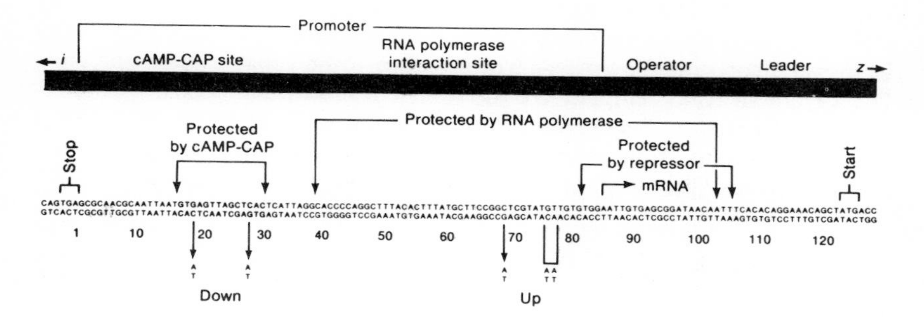

Fig. 4.8 The molecular structure of the regulatory region of the *E. coli* lac operon. Mutations which yield an inactive promoter are indicated as 'down' and those yielding a hyperactive or CAP site independent promoter as 'up'. (From Freifelder, *Molecular Biology*, Van Nostrand Reinhold, with permission).

of three adjacent protein coding genes, styled Z, Y and A, which code for beta-galactosidase enzyme, a permease enzyme, and a transacetylase enzyme respectively. These three proteins were found to be regulated coordinately. As shown in Fig. 4.8, the DNA containing these three gene sequences is preceded by other regions, known as i, p, and o, representing the regulatory gene, the promoter sequence, and the operator sequence. The last two are not, strictly speaking, genes, since they have no product, and should be referred to as DNA regions or sequences. The i gene codes for a single subunit of a tetrameric protein, the repressor protein, that has an affinity for the O region and binds to it. When O is occupied by repressor protein, no transcription occurs; in the presence of specific inducers, which, in the case of the lac operon, is normally allo-lactose, the repressor protein complexes with it and moves off the operator region. This is the result of a conformational change in the repressor when it is associated with inducer. Proteins which undergo such alterations in shape, changing the characteristics of one binding site depending on whether another binding site is occupied, are known as *allosteric proteins*. Activation of the gene by inducer is a process of derepression, allowing the RNA polymerase enzyme to move from the promoter region and commence transcription of the gene.

However, the lac operon also displays a different type of response since the attachment of the polymerase to the promoter region requires that the promoter also binds each molecule of cyclic AMP and a protein known as catabolite activator protein (CAP). The operon is therefore only active when cAMP and CAP occupy the promoter region and the operator region is free from repressors. These two aspects of regulation are referred to as positive (requiring activation) control and negative (requiring de-repression) control.

Operons are characteristic of bacterial gene organisation and regulation, but are not found in eukaryotes. However, the mechanics of regulation and the role of the allosteric gene regulatory protein produced by the i gene serve as important models of what may happen in the more complex world of higher

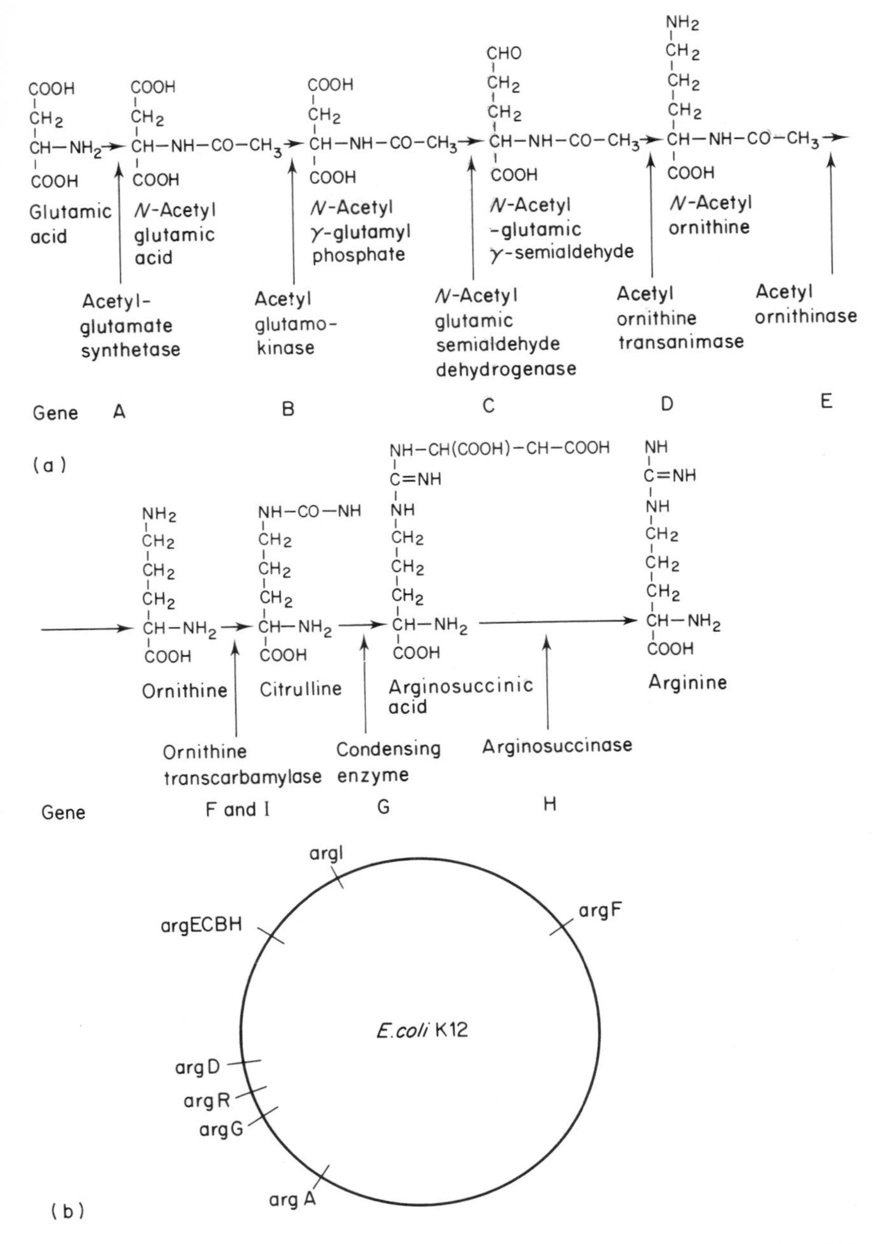

Fig. 4.9 The genetic system which is responsible for arginine biosynthesis in *E. coli*. The metabolic pathway is shown above, with the sites of action of the various enzymes, and the location of the relevant genes on the circular *E. coli* chromosome is indicated below. These genes together form a regulon. (From Lewin, B., *Gene Expression*, **3**, Wiley (1977), with permission).

organisms. Before leaving the prokaryotes, we should take note of a further elaboration of the operon structure, known as a *regulon*.

The conversion of glutamic acid to arginine in *E.coli* requires the assistance of eight enzymes, but the genes coding for these proteins map at eight distinct loci on the genome (see Fig. 4.9). In spite of their wide separation, all eight genes are expressed coordinately and are subject to repression by the product of one regulatory gene (together with the corepressor arginine) which is itself not closely linked to any of the eight genes. Such a series of separate loci under common control of one regulatory gene constitutes a regulon, and the arginine regulon presents a fascinating anticipation of gene regulation in eukaryotes, where, in the absence of true operons, gene regulatory molecules, such as steroid hormones, may modify the expression of many different genes widely dispersed in the genome.

4.5.2 Eukaryotic organisms use a range of mechanisms to regulate their genes

Prokaryotes, as we have seen, lack chromatin (and therefore any mechanisms of gene regulation based on such a complex) and depend chiefly on modulation based on positive or negative control of operons or single genes, such modulation being effected by proteins produced directly by regulatory genes. In studying gene regulation in eukaryotes, a much more complex picture presents itself. A great diversity of different mechanisms are in operation and different genes are regulated in quite distinct ways. Indeed it is clear that mechanisms of gene control have evolved more or less independently for different genes. It is also true that less is known about eukaryotic gene regulation than that of prokaryotes, which is understandable in the light of the elaborate changes in gene expression which accompany development and cell differentiation, and the relatively enormous numbers of genes at the disposal of the organism concerned. Therefore all that can be attempted here is a fairly brief account of the more important mechanisms.

Before addressing ourselves to examination of the details of individual mechanisms, it is appropriate to recap on the more general aspects already covered. These are that a series of three distinct RNA polymerases are used in the transcription of eukaryotic nuclear genes, that all genes have somewhat similar promoter sequences that serve to attract and bind the polymerase, and that specific start and stop signals serve to limit the transcript precisely. Also that some genes, such as those of ribosomal RNA, may well be expressed constitutively, while many others, especially those coding for tissue-specific products, are expressed only in very special circumstances. In addition, all eukaryotic genes are complexed with histone to form chromatin, and present evidence favours the view that with certain possible rare exceptions (the ribosomal RNA genes might be one such), the histones and their nucleosomal aggregates persist whether or not a stretch of DNA is being transcribed (though perhaps not at the promoter). We have also noted that many eukaryotic genes are interrupted by introns, which are expressed in the original product of transcription, the high molecular weight hnRNA, and that the mRNA is, in most cases, the result of intron removal and exon splicing within the nucleus. All of the above, are common features which are necessary for transcription and expression, but do not, for the most part, determine whether or not a gene is to be expressed. In other words, they are not selective mechanisms. To discover such selectivity we must now look at certain other factors which affect eukaryotic transcription.

4.5.3 Chromatin conformation

When the polytene chromosome was discussed in Chapter 3, the point was made that chromatin that was transcriptionally active, whether in the decondensed puffs of the bands, or the constitutively expressed sequences of the interbands, was always in an open extended conformation as compared to the highly compacted and transcriptionally inactive chromatin of the chromomeric bands. It will therefore come as little surprise that this broad correlation between transcriptional activity and an extended chromatin conformation is a rather general aspect of eukaryotic gene regulation.

A most informative experiment in this area was carried out in 1976 by Weintraub and Groudine. Since it was already clear that chromatin, when decondensed, was readily digested by the endonuclease enzyme DNase I, these authors combined such digestion with the then recent availability of radioactively labelled DNA probe sequences. The experiment they carried out was as follows. Chromatin was extracted from various tissues, including oviduct and reticulocytes, of the domestic chicken. This chromatin was then exposed to a set concentration of DNase I under conditions where it was known that extended chromatin would be preferentially digested. Following such digestion, DNA was purified from each sample and exposed in appropriate hybridizing conditions to radio-tagged cloned DNA sequences of chicken globin and ovalbumin genes. The findings were that the globin gene would hybridize with the DNA from oviduct, and the ovalbumin gene with DNA from reticulocytes, but hybridization in the complementary situation was not detected. This strongly suggested that the globin gene had been preferentially digested in reticulocyte chromatin and the ovalbumin gene in oviducal chromatin, presumably because these genes in these tissues were specifically and extensively in locations of decondensed chromatin.

Some further observations can be added to this general conclusion that active genes are to be found in decondensed chromatin. The first is that chromatin may be opened up prior to transcription commencing, especially when developmental stages involve the activation of say, beta globin synthesis at a set time. In such a situation the decondensation may precede the transcription by hours or even days. So it seems that the alteration in conformation is a necessary but not sufficient condition for gene activity. A second observation is that the actual length of decondensed material may be much greater than the strict length of the sequence to be transcribed. We noted that a polytene chromosome puff frequently involved an entire chromomere, which has enough DNA for some twenty genes. So it may be that long flanking regions up and down stream from the coding sequence must also be involved in the activation process. Finally, it has been found that some regions in the promoter region of an active gene become extremely sensitive to DNase I shortly before transcriptional activation of a sequence. The presence of such hypersensitive sites has been noted in connection with activation of many genes (Elgin, 1981), and is suggestive of the localised loosening or loss of nucleosomes from confined regions of chromatin.

The mechanism of chromatin decondensation, except when it involves actual loss of nucleosomes, is probably a function of the differential activity and binding characteristics of histone H1. As we noted in Chapter 3, this histone partially straddles the gap between neighbouring nucleosomes and there is quite a lot of circumstantial evidence that it fulfils a key role in chromatin condensation and decondensation (Butler and Thomas, 1980) However, as we shall see in the next two sections, H1 is not the only factor to mediate changes in chromatin conformation.

4.5.4 Histone acetylation and ubiquination

It has been known for some time that the chromatin in which transcriptionally active gene sequences reside has histones which are highly acetylated. Such acetylation is a reversible modification to the histones, especially of the core histones, and converts lysine residues to acetyl lysine. All the sites of histone acetylation are located in the basic N-terminal regions of the molecules which tend to protrude slightly from the nucleosome surface, and the net effect of such acetylation is to neutralise the net basic charge at the histone extremities. Although the correlation between histone acetylation and the transcriptional activity of chromatin is quite striking, (Allfrey, 1977) it remains unclear whether the acetylation is necessarily causitive of gene activity, that is, whether

10
Met – Gln – Ile – Phe – Val – Lys – Thr – Leu – Thr – Gly –
20
Lys – Thr – Ile – Thr – Leu – Glu – Val – Glu – Pro – Ser –
30
Asp – Thr – Ile – Glu – Asn – Val – Lys – Ala – Lys – Ile –
40
Gln – Asp – Lys – Glu – Gly – Ile – Pro – Pro – Asp – Gln –
50
Gln – Arg – Leu – Ile – Phe – Ala – Gly – Lys – Gln – Leu –
60
Glu – Asp – Gly – Arg – Thr – Leu – Ser – Asp – Tyr – Asn –
70
Ile – Gln – Lys – Glu – Ser – Thr – Leu – His – Leu – Val –

Leu – Arg – Leu – Arg

Fig. 4.10 The nucleotide sequence of calf thymus ubiquitin. (Reproduced from Schlesinger, D.H. *et al.*, *Biochemistry*, **14**, 2214 (1975).

the modification precedes or follows from the changes in chromatin conformation alluded to in Section 4.5.2. A similar question arises over the preferential association of the protein ubiquitin (see Fig. 4.10) with the histone H2A (a complex referred to as A24 protein) in the chromatin of active genes (Levinger and Varsharsky 1982). Is it, at least partially, a cause of chromatin decondensation, or simply a result of the greater availability of H2A following such decondensation?

4.5.5 HMG proteins in chromatin

HMG proteins (high mobility group proteins) are relatively small, highly charged proteins that migrate close behind the histones on gel electrophoresis of chormatin protein. Two such proteins named HMG 14 and HMG 17, are frequently found to be enriched in active chromatin, each nucleosome bead being capable of binding two molecules of either HMG protein. The evidence suggests a structural role for these proteins in chromatin, but, as with histone acetylation and ubiquination, a causitive role in chromatin decondensation is possible but less clear.

4.5.6 DNA methylation

While proteins such as histones can undergo a range of post-synthetic modifications – acetylation, phosphorylation, and methylation – the DNA undergoes only one, namely DNA methylation. It involves either the conversion of cytosine to 5-methylcytosine or of adenine to 6-methyladenine. Since the latter type of methylation is confined to prokaryotes (and a few specialised organisms and genes in eukaryotes), we will not consider it further, other than to emphasise that it serves a role in bacteria not of helping to regulate genes but of protecting genomic DNA from the nuclease enzymes that break down invading phage DNA.

Methylation of cytosine residues in eukaryotes is irreversible and normally affects GC pairs. When such a sequence is replicated, the new and unmethylated strand will have the complementary sequence CG, and the new C residue is then methylated in this partially symmetrical fashion. If the methylation of DNA is a growing tissue is to be reduced, the only way in which this can be accomplished is by blocking this symmetrical methylation of the new strand, so leaving a hemimethylated DNA duplex.

The relationship between DNA methylation and gene regulation in eukaryotes is complex (see review by Bird, 1983), but there is a marked tendency for heavily methylated gene sequences to be inactive. This negative correlation between gene activity and DNA methylation is far from constant, but the correlation seems much stronger if the methylation of C residues in the promoter region of the appropriate gene sequences is considered. It is as if methylation of certain specific promoter sites may inhibit the proper association with the RNA polymerase enzyme and thus reduce the likelihood of transcription.

The most crucial evidence that DNA methylation might be a cause rather than a result of gene inactivity comes from experiments employing the drug 5-azacytidine. This is a base analogue which is incorporated into the DNA of dividing cells and, when incorporated, prevents the following methylation of the cytosine residues. During some remarkable experiments in which this drug has been introduced into primates (including the human), inactive genes such as that for gamma globin (normally heavily methylated and also inactive in the adult, but unmethylated and active in the foetus) have been preferentially activated. Patients suffering from beta thallasaemia could therefore be rescued from the effects of the disease. However, considerable caution must be exercised in interpreting these experiments, since 5-azacytidine is also a cytotoxic drug and could be affect gamma globin synthesis by its tendency to damage the liver, a site of synthesis of blood cells in the foetus.

An important objection to the idea that DNA methylation is a major mechanism of gene regulation in eukaryotes is the observation that the DNA of some organisms, such as that of the fruit fly *Drosophila*, is not methylated at all. So many unanswered questions remain in this area of investigation.

4.5.7 Gene regulatory proteins

Many proteins regulate genes in unspecific ways but here we are concerned with proteins that have a role in the specific activation or repression of individual genes at particular times in development or in particular tissues. Whenever the role of the repressor protein in regulation of the lac operon of *E. coli* (see Section 4.5.1) became clear, the search began for proteins with analagous functions in eukaryotes. Although the evidence was slow in coming, there are now a few well authenticated examples. These we will consider in turn.

A transcription factor protein for the 5S genes

The small ribosomal RNA species of eukaryotes is a 5S molecule, and there are hundreds, even thousands, of copies of these sequences in some organisms. A remarkable discovery was made in relation to their transcription by Bogenhagen *et al.* (1980), namely that deletion of a 40 base pair region from the centre of the coding sequence obliterated all transcription. In subsequent experiments it transpired that the RNA polymerase III molecule did indeed use this central intragenic sequence as a promoter, but that it could only do so when an additional protein was involved in the interaction. This protein, now styled transcription factor III (TF IIIA), has been isolated and proved to be effective in specifically activating expression of the 5S sequence *in vitro*. (See discussion by Miller, 1983). Another interesting finding about this protein is that it also binds specifically to the 5S RNA itself, so permitting the accumulation of this species of RNA in a cell to inhibit its own synthesis by competition for TF IIIA. In other words, a form of end product inhibition of synthesis is mediated by this protein.

Heat-shock protein in *Drosophila*

All organisms are likely to suffer periodic effects from the shock of elevated temperature and many have specific proteins synthesised shortly after heat-shock: these presumably have a role in protecting the cell from the more deleterious effects of the high temperature. It has long been known that specific heat-shock puffs can be detected in the polytene chromosomes of *Drosophila* when the salivary glands bearing these chromosomes are heat-shocked. Presumably these heat-shock puffs are chromosome areas in which the genes responsible for heat-shock protein production are located, and are therefore involved in preferential transcriptional activation. In 1978 Compton and McCarthy capitalised on this interesting situation by exposing isolated polytene nuclei from *Drosophila* larval salivary gland cells to cytoplasm from *Drosophila* cells heat shocked in tissue culture. These workers were able to demonstrate that such cytoplasm had the effect of specifically inducing the appearance of heat-shock puffs in the chromosomes of such nuclei. Subsequent work has shown that proteins can be isolated from such heat-shocked cytoplasm which have the capacity to induce heat-shock puff formation. These proteins seem good candidates for the title and role of specific gene regulatory proteins.

The proteins of homeo-box genes

A remarkable story has unfolded in the last few years in relation to genes in *Drosophila* which are involved in the so-called homeotic mutations. These mutations lead to adoption by one segment or part segment of the fly of a morphological form normally characteristic of a different segment. For example, wings may come to replace halteres or legs to replace antennae. When the techniques of gene cloning and sequencing came to be developed, it was discovered that all of these homeotic genes had a common consensus region within their coding sequences, the so-called homeo-box. Almost overnight, many different laboratories around the world began a search in other organisms for genes which contained the homeo-box region, and sure enough, such sequences were identified. A series of papers on the topic of the homeo-box genes can be found in the journal *Cell*, Volume **43** (1), 1985. Initially it was suspected that genes with homeo boxes were sequences involved in determining segmentation but this view has now been superceded by the conclusion that they are essentially genes which regulate other genes during development, and that the homeo-box region is itself a sequence which codes for a protein domain which functions by binding to DNA. Even gene regulatory proteins in bacteriophage lambda have regions homologous to homeo-boxes. The homeo box genes in *Drosophila* are often expressed in development at sharply defined times, and in specific parts of the embryo, and so have many of the characteristics expected of genes whose protein products regulate other genes important in developmental gene expression.

4.5.8 Other molecules with gene regulatory functions

Proteins are not the sole intermediates in modulating the selective gene expression which characterises eukaryotic organisms. It would be surprising if RNA does not also serve this function, at least in certain defined situations. But another class of substances, the steroid hormones, have long been known to have quite specific effects on gene expression. Experiments have focussed on two steroids, oestrogen in its effect on oviduct tissue, and ecdysone as a developmental hormone in insect development. What is now quite clear, and the polytene chromosome comes to the forefront as the experimental situation of choice, is that ecdysone does bind to specific gene sequence locations in chromatin and is capable of inducing the puffing of the genes with which it associates. So once more we have good evidence for specific gene activation by a molecule whose role is the induction of particular biochemical changes in cells and tissues.

4.5.9 Nucleus and cytoplasm

In the next chapter consideration will be given to how cells become committed to different fates and how nucleus and cytoplasm are, together or separately, involved in the decision-making process. Aside from the interesting question of cell commitment, it is obvious that in a eukaryotic cell, both nucleus and cytoplasm must communicate and that in most cases the primary signals which indicate the requirement for a change in gene activity are encountered first by the cell surface and/or the outer cytoplasm of the cell. After all, it is most frequently in response to changing circumstances around it that a cell will alter its pattern of gene expression. How is this achieved?

The question is a complex one, partly because, as has been already stressed, different genes are regulated in different ways, but even more because nucleus and cytoplasm are constantly exchanging information. All proteins which help regulate the DNA, polymerases, histones, HMG proteins, transcription factors and other gene regulatory molecules are all synthesised in the cytoplasm and no doubt the cytoplasm can modulate carefully their availability to the nuclear genes. On the other hand, the cytoplasm cannot go far without the nucleus since protein can only be made at the dictates of the mRNA which originates in the nucleus, and the tRNA and rRNA which are the cytoplasmic workhorses of the translational apparatus are synthesised purely in the nucleus. So the ground rules all underline interdependence between nucleus and cytoplasm at every level. However, we can categorise a number of distinct patterns which will all contribute to the total scenario which enables a cell to functionally regulate its genes.

(a) Signals will be received by the cell surface, either as altered ionic fluxes outside the cell, or distinct identifiable molecules such as polypeptide hormones settling on the surface. This information will be relayed to the nucleus, often after cytoplasmic 'filtering', and the genes will be turned on or off accordingly.

(b) Molecules, such as steroid hormones, will enter the cell via specific cell surface routes and will be conducted to the nucleus under strict cytoplasmic control. There they may interact with specific genes directly to accomplish appropriate changes in gene activity.

(c) Some changes in the transcriptional pattern are probably accomplished in response to purely nuclear events. Thus the activity of one gene can produce an RNA molecule which may trigger another into action or inaction, and the nucleus may change its activity pattern in response to elapsed time or numbers of mitoses.

(d) The pattern of gene activity is partly regulated by specific proteins which are themselves the products of a regulatory gene, but must be made in the cytoplasm and from there return to the nucleus to help control other gene sequences. Here the cytoplasm may exert an influence by modulating the translation rate of the specific regulatory protein or its effective passage to the nucleus.

Taken together, gene regulation is a cellular phenomenon rather than a purely nuclear one, and there is almost no end to the intricacies of the controlling networks.

4.5.10 Concluding remarks on gene regulation

This chapter does not in any sense represent an exhaustive account of what is known about gene regulation in prokaryotes and eukaryotes, but an attempt has been made to select the more important or better understood mechanisms. As was stated at the outset of this chapter, different genes and gene families almost certainly use different mechanisms or different sets of mechanisms to achieve the same regulatory ends. And since DNA methylation and histone acetylation have a key role in regulating one sequence there is no reason for supposing that they have an identical role in regard to some other sequence.

In general terms an understanding of gene regulation must distinguish between general factors and specific factors. The RNA polymerase enzymes provide general regulation and in their absence transcription cannot occur. But heat-shock protein is an example of a specific factor which is vitally important in the control of a few genes related to the heat shock response. It is also useful to distinguish between so-called cis-acting sequences and trans-acting sequences (a terminology which derives from the cis/trans test of classical genetics). Cis-acting factors are sequences or products of sequences which are on the same chromosome and indeed on the same DNA molecule as the gene affected by them. Thus enhancer sequences are cis-acting and probably act directly and not via an intermediate product, while many gene regulatory sequences which are effective through protein intermediates, such as the heat-shock or homeobox genes of *Drosophila* are probably all trans-acting, in that the genes affected are not linked to the regulatory region and are probably situated on a different chromosome.

Many of the most fascinating aspects of eukaryotic gene regulation are those that are intimately involved in selective gene expression, an aspect of genetics that is itself bound up with cell differentiation. In the following chapter we will pursue further our consideration of gene regulation in the context of its phenotypic expression in the cell and in the organism.

5

Gene Regulation and Cell Differentiation

5.1 Cell differentiation and selective gene expression are correlated

Although this book is chiefly concerned with genes and their regulation, there are good reasons for including a chapter largely devoted to the topic of cell differentiation. The main reason for this is that it is within the context of this biological phenomenon that gene regulation is supremely important, but also as a result of the tantalising chicken and egg situation presented by these two topics. Is gene regulation in itself a sufficient explanation for the mechanics of differentiation, or is gene regulation simply an accompaniment and result of this latter process, the crucial determinative aspects of cell specialisation coming from outside the genome and its inner controls?

Curiously this circularity of argument regarding differentiation and regulation which is required by restricted genetic expression is often encountered in text books and monographs relevant to these topics, so it seems all the more necessary to discuss their relationship here. An examination of this topic in greater depth will be found in Maclean and Hall (1987).

Differentiation is the process in which different cells in a multicellular organism become specialised to fulfil distinct functions. Since, within one organism, all of these different cells retain a complete genetic complement (with some very rare exceptions such as the chromosome diminution found in the worm *Ascaris*), it is quite obvious that cell specialisation is not a result of selective gene loss. Nor is this specialisation in the main a result of manipulation of pre-existing populations of proteins or RNA, although such post-transcriptional regulation involving differential degradation rates for different RNA and protein molecules does occur. Rather, it is clear that the weight of the evidence points towards the level of RNA synthesis as the crucial way in which cell specialisation is engineered molecularly. Therefore by expressing only a restricted subset of the entire genome, a cell ensures that only certain proteins out of the total range are made, and these in varying amounts, so that a cell is produced which, enzymatically and structurally, is clearly differentiated from other such cells in the same organism which are adapted to different specialisations.

It is also noticeable that cells specialised in the same way are strikingly similar both biochemically and morphologically. Thus a complex multicellular organism, such as the human, has somewhere between 150 and 200 different specialised cell types (assuming that different memory neurones and different antibody-forming lymphocytes are not, strictly speaking, cells of different types). Cells of any one cell type within such an organism, say liver parenchyma cells, are strikingly similar to one another. Indeed if cell cycle stages and chronological age variation is allowed for, such cells can probably be usefully considered to be identical. So by operating precise gene subsets in about 200 distinct patterns, a complex eukaryotic organism can engineer the intricate pattern of cells which go to form its organs and tissues. This helps to explain why in any one cell at one point in time a very restricted portion of the genome is found to be active. All of the genes appropriate to other distinct and separate cell types remain silent.

Notice that we have not concluded that selective gene expression explains cell differentiation, for we have not discussed how a cell comes to choose the appropriate subset of genes relevant to its destiny. Rather, it has been emphasised, selective gene expression is the mechanism whereby a chosen specialised cell fate is accomplished in space and time.

Since this strong correlation between cell differentiation and selective gene expression forms the cornerstone of this chapter, it is appropriate to review briefly the evidence that supports it. Ability to believe in something deeply is, after all, greatly dependent on knowing and understanding the evidence that underpins it. I will cite three independent items of evidence which support the correlation.

1. the evidence from electron microscopy of spread films of chromatin from specialised eukaryotic cells that most of the DNA is transcriptionally inactive,

2. the evidence already discussed in Chapter 4 that genes are amenable to DNase 1 digestion in tissues in which their products are found but not in tissues lacking such products, and

3. that the puffing pattern of insect polytene chromosomes is cell type specific. This latter evidence is probably the most visually satisfying of the three, since, if tissues such as salivary gland, midgut and malphigian tubule are compared in a *Drosophila* larva, all will be found to contain polytene chromosomes, and the basic chromomere banding pattern in all three will be identical. However, the pattern of puffing will vary from tissue to tissue, but not from cell to cell within one cell type in one tissue. Now since puffing is a visible expression of intense transcriptional activity, it is clear that such patterns of activity are strongly correlated with cell differentiation.

Having now established the close relationship between cell differentiation and the orchestration of restricted sets of genes within the genome, let us proceed to examine some general aspects of differentiation before attempting, at the close of the chapter, to determine what, in terms of gene regulation and differentiation, is the primary phenomenon. In other words, what is and what is not causative. It is customary to refer to the chosen specialisation of differentiated cells as cell commitment, this terminology implying that the choice of fate may be distinguished from the gradual unfolding of such specialised fate, which is the definition of differentiation. Five distinct aspects of commitment are discussed below and each aspect will then be considered in turn.

Important aspects of cellular commitment

(a) It is often divisible into distinct stages, termed determination and differentiation.
(b) It is normally very stable.
(c) Its stability is cellular rather than nuclear in origin.
(d) It is often progressive in onset.
(e) Its stability is not absolute in all cases.

5.2 Cell commitment can often be divided into distinct stages, namely determination and differentiation

The twin stages of cell commitment can be distinguished in some, but not all types of specialised cells. The first stage, determination, can only be recognised

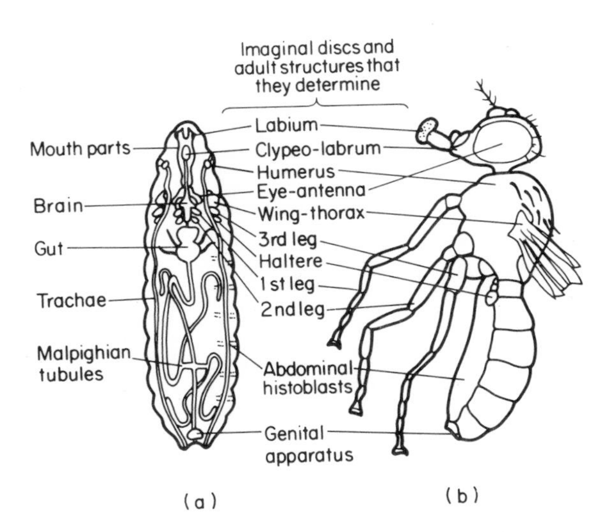

Fig. 5.1 Imaginal discs of a *Drosophila* larva and the adult structures which they determine. Larval organs are indicated on extreme left, larval discs on right of the larva, and adult structures arising from discs on left of the adult fly. (Reproduced with permission from Sang, *Journal of Genetics and Development*, Longman (1984)).

experimentally, since it is a state of fixed potential without overt signs of specialisation. The biological situation that illustrates the distinction between these two situations most clearly is that of the imaginal discs, the small nests of cells which, in holometabolous insects such as *Drosophila*, are the embryological precursors of appendages such as wings, legs or genitalia, as seen in Fig. 5.1. These small groups of cells can be found as hollow pockets invaginated below the epidermis of the larvel insects. On microscopic examination the cells can be seen to be unspecialised and embryonic in character, and do not differ from disc to disc. The hormonal environment of the larva prevents their further development, but in the pupa, where the hormonal climate is different, each disc everts and grows dramatically, the cells rapidly becoming differentiated in line with their appropriate fate. But the crucial observation is that even the undifferentiated cells of the larval discs can be shown experimentally to be committed to their cellular destiny, in other words to be determined.

The experimental evidence for imaginal disc determination stems from a remarkable experimental technique. As shown in Fig. 5.2 it is possible to

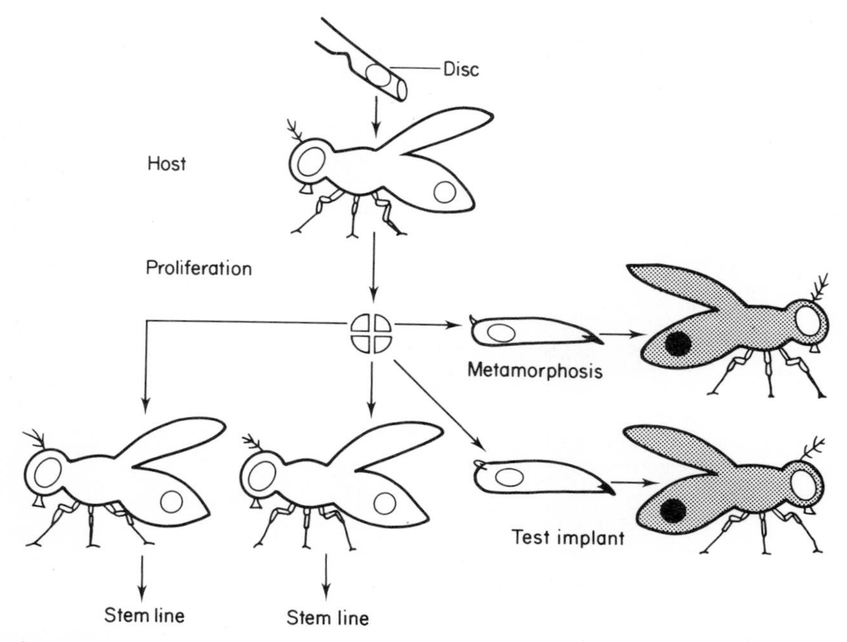

Fig. 5.2 The process by which imaginal discs may be passaged in the abdomens of adult *Drosophila*. Individual discs may be cut into fragments and some fragments of a disc repeatedly passaged, while other fragments of the same disc are placed in larvae to test their determined fate. Although transplanted discs do not grow into normal organs or appendages because of their altered location, the characteristics of the structure produced are sufficient to permit unambiguous identification. (Reproduced with permission from Sang, J. Genetics and Development, Longman (1984) – modified from Hadorn, E., Developmental Biology, **7**, 617–29 (1963)).

remove a disc from a larva and implant it into the haemocoel of an adult fly, where it survives but does not differentiate. Indeed one disc can be passed many times from adult to adult, thus becoming chronologically aged but the cells remaining in their original embryonic form. If, after many such passages, the disc is now once more inserted into a larva, during subsequent pupation the differentiated form of the disc may be discovered by dissection and microscopic examination. Of course the disc cannot form an entirely normal organ or appendage, but experienced biologists have little difficulty in establishing the identity of the differentiated tissues. It has been shown many times that such artificially subcultured discs retain their fate memory, indicating that the original disc, though not overtly differentiated, was indeed determined.

5.3 Cell commitment is normally very stable

We have just seen that determination, as exemplified by the insect imaginal disc situation, is very stable. So too is overt differentiation. When explants of a tissue such as kidney or skin are established in tissue culture, the constituent cells of that tissue normally retain their separate identities, even after long periods of continual growth in culture. Thus cells such as cardiac muscle in culture are still readily distinguishable as such more or less indefinitely; they remain capable of coordinated contraction in culture, retain a characteristic morphology, and will be found biochemically to contain and synthesise proteins characteristic of such cells. So commitment is, in most cases, a state of permanent dedication to a specific fate. It is as if, once the switch has been thrown that indicates the choice, reversal is in most cases impossible. However, as we shall see section 5.6, the stability of commitment is not absolute and some cases of significant directional change can be identified. However, they remain very much the exception. It must also be emphasised that the stability of commitment persists through many rounds of mitosis, so that, whether *in vivo* or *in vitro*, once committed, even the trauma of cell division, with its concomitant chromatin condensation, transcriptional shut down, and eventual nuclear division, does nothing to alter the strict conformational specialisation of the cell in question.

5.4 The stability of commitment is dependent on both nucleus and cytoplasm

Since microsurgery and artificial cell fusion offer the possibility of exchanging nuclei and cytoplasm between cells, it is clearly possible to address an interesting and important question. Is the stability of commitment a purely nuclear or purely cytoplasmic phenomenon? An excellent monograph by Harris (1970) summarises much of the less recent evidence. Original attempts to answer this question exploited the properties of a remarkable unicellular alga called *Acetabularia*. This gigantic cell, some 3–5 cm in height, is a common Mediterranean organism of the subtidal zones. It has a basal rhizoid in which the single large nucleus is located, and an elongated stem rising from the rhizoid. This stem is topped by a hat-shaped cap, which varies in form depending on species, and contains the numerous sexual gametic cells. In the early 1960s Hammerling (1963) carried out microsurgery with this organism, exchanging rhizoids (with nuclei) and fruiting stems between species. By removing the cap from the stem, he was able to determine whether a nucleus of a different species, grafted on to the base of a stem, would affect the form of the new apical cap. What happened was that the new cap closely resembled its predecessor; when it in turn was removed, and a second regenerate was also dissected away, a third regenerate cap forms; this third cap is of a type characteristic of the species donating the basal nucleus. It seems evident that once the store of long-lived messenger RNA in the cell cytoplasm is exhausted, the nucleus is able to dominate the morphology of the new cell structure. So this early work on *Acetabularia* seems to emphasise the primacy of the nucleus as the determinant of cellular morphology.

However, recent experiments provide a different conclusion. These involve the technology of transplanting the nucleus of a somatic cell into a fertilized egg, the egg nucleus having been previously destroyed by exposure to ultraviolet irradiation. Donor nuclei can be recovered from specialised cells in order to ascertain whether the egg cell will then become a cell with a similar specialisation. Many different laboratories have carried out this type of experiment, usually with eggs and cells from various species of amphibians. Although not all such experimental programmes have been successful in growing somatic cell nuclei/enucleated egg fusions into adult animals, most have been able to grow them on to larvae (see review by Etkin and Diberadino, 1982). Some of the most dramatic and successful experiments have been carried out in John Gurdon's laboratory (see Gurdon and Mellor, 1981). The experimental protocol involved recovering intact nuclei from such conspicuously differentiated cells as cultured amphibian skin cells, shown to contain the specialised protein keratin, which is characteristic of such cells. In earlier experiments ciliated gut epithelial cells of swimming tadpoles had also been used. Such nuclei were injected singly into enucleated eggs, as shown in Fig. 5.3, and development allowed to proceed. A proportion of these eggs developed into larvae, complete with blood, cartilage and nerve cells, and in some cases such larvae produced adults. This technique has been used by Gurdon's laboratory to produce a clone of albino adults, see Fig. 5.4.

These remarkable experiments not only demonstrate convincingly that the nuclei of many differentiated amphibian cells are genetically intact, but also that differentiation does not necessarily constrict the potential genetic expression of the nucleus. When the nucleus is removed from its original cytoplasmic environment and introduced into a quite novel one, an entirely new repertoire of expression results. This should not be taken to imply that the cytoplasm alone is the mysterious custodian of differential fate, but rather that the nucleus alone is insufficient and therefore that differentiation is a cellular phenomenon, requiring a complex series of interactions between nucleus and cytoplasm.

Since the above experiment has been cited as demonstrating nuclear

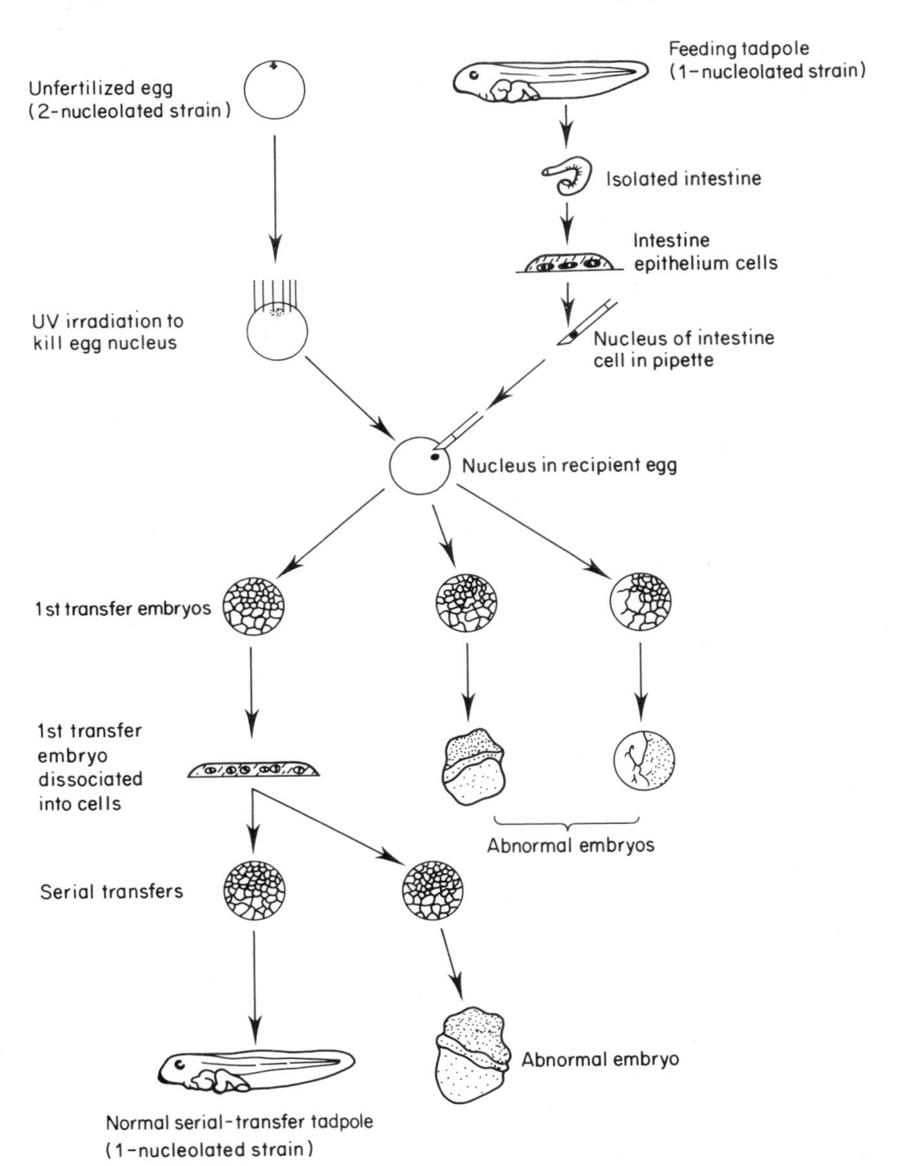

Fig. 5.3 Protocol for transplantation of nuclei from intestinal epithelial cells of *Xenopus* larva into unfertilized eggs in which the egg nuclei have been destroyed by ultraviolet irradiation. Some develop into normal tadpoles, others into abnormal and short lived embryos. (Reproduced with permission from Graham, C.F. and Wareing, P.F. (ed), *The Developmental Biology of Plants and Animals*, Blackwells, (1976)).

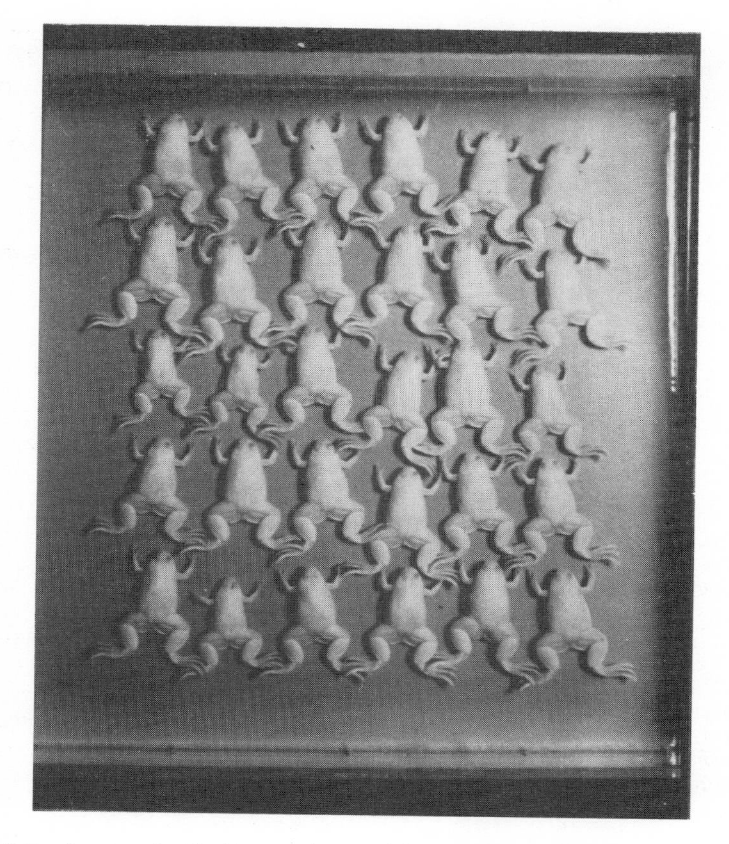

Fig. 5.4 A series of white *Xenopus* laevis adults which form a genetic clone. All derive from transplantation of single somatic nuclei from a white individual into a series of UV irradiated eggs from a normal adult. (Photograph kindly provided by Dr. J. Gurdon).

totipotency, it is well to emphasise that not all specialised cells have proved as amenable in supporting development. Rather similar experiments have been attempted in mammals, but in this case only a few rounds of division result and no cell specialisation in the embryo can be achieved. In this situation it is difficult to determine whether purely technical obstacles are involved and ultimately even mammalian nuclei will prove capable of complete reprogramming, or whether some real barrier to future potential expression is an essential characteristic of the specialised mammalian cell.

These two sets of evidence from *Acetabularia* and amphibian eggs actually complement one another. We can learn from *Acetabularia* that a new nucleus can redirect the cytoplasm, and from the work on amphibian egg that new cytoplasm can persuade a nucleus to alter its pattern of transcriptional activity, and thus to change its differentiative commitment. Both nucleus and cytoplasm are closely interdependent in a eukaryotic cell. The mRNA that is made in the nucleus migrate to the cytoplasm and is there translated into protein. But many of the cytoplasmic proteins are known to re-enter the nucleus as important constituents of chromatin, as are the histones, or as specific gene regulatory molecules, as are some of the proteins discussed in Chapter 4. So the specialisation which characterises a differentiated cell is evident both in the instructions passed from nucleus to cytoplasm in the form of mRNA and instructions passed from cytoplasm to nucleus in the form of specific gene regulatory proteins. If either nucleus or cytoplasm are changed, the stability of commitment is lost, at least temporarily, and often permanently.

5.5 Commitment is often progressive in its onset

Earlier in this chapter we have discovered that many cells, en route to final specialisation, pass through a stage of determination. At this phase of commitment, although overt signs of morphological specialisation are absent, experimental manipulation reveals that the choice has already been made, the path to final cell specialisation chosen. But presented in this light, it might be assumed that in no case was there any sign of the onset of differentiation in any determined cell, and that the onset of differentiation was sudden and total. Such is rarely the case. More often a progressive specialisation is evident, and only the cell at the terminus of the process is equipped in every sense for this final specialisation, whether as a neuron, lymphocyte, chondrocyte or epithelial cell.

In many cases the progression to specialisation will involve a number of rounds of cell division and the onset of specialisation may be seen morphologically or biochemically in some cells prior to the final mitosis. It is often possible to utilise sensitive assays for cell specific proteins that characterise certain differentiated cell types. For example, fibroin will characterise silk gland cells, keratin some epidermal epithelial cells, immunoglobulin certain lymphocytes, crystallin the lens fibre cells, and haemoglobin the mature erythrocytes. Using radio immune assays or ELIZA systems which depend on colorimetric detection based on immune conjugates, traces of these specialist proteins may be found in cells which are intermediates on the pathway to specific specialisation. Detection of even earlier signs of specific gene activity can utilise identification of particular species of messenger RNA by hybridization with radiolabelled complementary DNA probes. Such an assay has been used in Fig. 5.5 to detect globin mRNA in various types of blood cells differentiating in culture, and it will be evident that this specialist RNA species is not confined to the mature erythrocytes, but is found in lesser amounts in earlier precursor cells. Indeed it can often be demonstrated that the mRNA for a specialist protein is detectable some time before the appearance of the specific protein whose amino acid sequence it dictates.

In order to illustrate the progressive nature which typifies many examples of cell differentiative systems, we will look more closely at one such system, namely erythropoiesis or blood cell formation. Admittedly, in this system commitment is more progressive than in most but it does illustrate the prin-

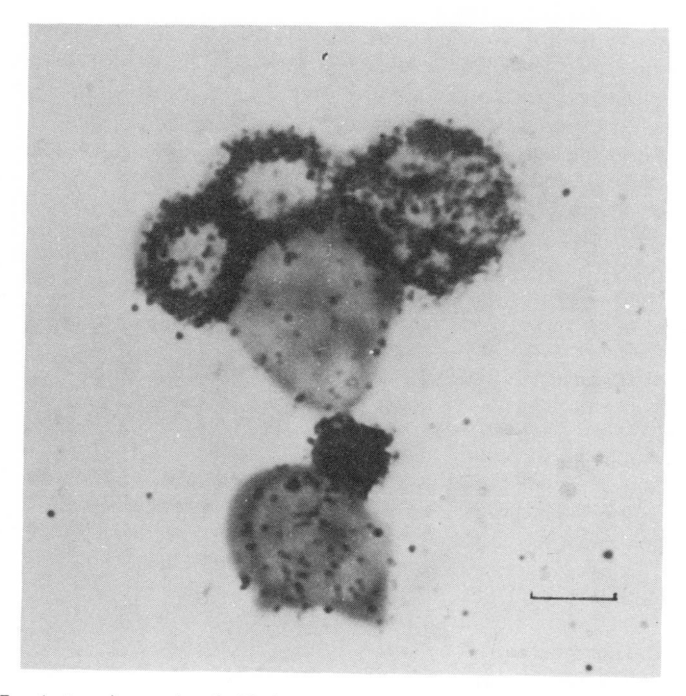

Fig. 5.5 Autoradiograph of 13-day mouse foetal liver cells exposed to *in situ* hybridization with tritium-labelled globin complementary DNA. The grains are therefore indicative of the presence of globin messenger RNA. The small heavily labelled cell is a reticulocyte, the two labelled cells of intermediate size are basophilic erythroblasts. The largest cells are proerythroblasts, of which one is labelled, one is slightly labelled, and one is not labelled above background. Scale bar 10 μm. (Photograph kindly provided by Dr. P. Harrison).

ciples of such progressive onset in a particularly clear manner. The biological situation we will consider is that found in mammals, but it is probably broadly similar in most vertebrates.

The steps and processes that are involved in blood cell formation are referred to as haemopoiesis or erythropoiesis, despite the fact that many of the cells produced are neither erythrocytes nor synthesise haemoglobin. Indeed a number of quite distinct blood cell types are generated, for example, erythrocytes, lymphocytes, granulocytes and megakaryocytes. The latter cells later break up into small fragments to yield the blood platelets that are important in blood coagulation. All the distinct cell types included under the blood cell umbrella are the products of one original stem cell type, the so-called multipotential stem cell. Notice that this stem cell is already potentially committed in that it can produce only blood cells but it is multipotent in that blood cells of many kinds may be generated from it. So unlike, say, a fertilized egg, it is not obviously totipotent, in that it has some restriction placed on its capacity for genetic expression. Yet this restriction is not a result of gene loss, and no doubt if its nucleus were placed in a new cytoplasmic environment, a much broader range of possible genetic programmes would be revealed. These multipotential stem cells of the erythropoietic cell line are set aside from other cells in early embryonic life, at or before, the time of yolk sac development, and no new stem cells seem to be produced at later stages of development. This explains the extreme difficulty of treating patients with aplastic anaemia, since in this disease a net loss of these multipotential stem cells occurs.

As shown in Fig. 5.6 the cell divisions occurring in this stem cell population do not yield the final products of the various differentiative pathways. Instead, a series of other stem cells are produced. but these now have a further commitment, so that they can only produce cells specialised for one, or at most two (in the case of macrophage and granulocytes) of the differentiative pathways. Notice also that these stem cells are also self generating and so have mitotic choices of producing more of themselves, or cells more narrowly committed than themselves, or both. These various alternatives are known to be open to

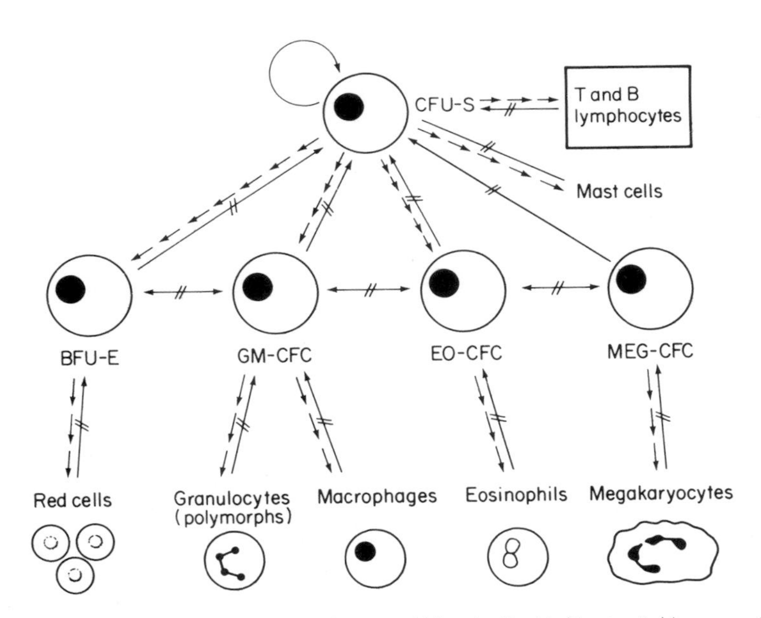

Fig. 5.6 The early events in the production of blood cells. Multipotential haemopoietic stem cells, i.e. colony-forming units (CFC-S) are capable both of self replication and the generation of a variety of specific progenitor cells by a sequence of proliferative and differentiative events. Progenitor cells cannot dedifferentiate or transform to other progenitors but are able to generate clones of maturing progeny cells, the most mature of which appear in the blood. Note that many granulocyte-macrophage progenitors, i.e. colony-forming cells (GM-CFC) are bipotential and able to form both granulocytes and macrophages. BFU-E erythrocytic burst forming unit. (Reproduced with permission from Metcalf, D. in *Cellular Controls in Differentiation*, C.W. Lloyd and D.A. Rees (eds), 125–47, Academic Press (1981)).

hormonal and other influences, thus permitting considerable plasticity. Therefore, if drastic blood loss due to haemorrhage has occurred, the stem cells can switch major productivity to the erythroid pathway, and economise temporarily on making more stem cells or cells of other blood cell categories. There is a progressive aspect to commitment in the form of a developmental cascade, with the multipotential stem cell at the top of the cascade and the various 'end cells' such as lymphocytes, themselves now incapable of further division, at the bottom. There is also a further aspect of progressive differentiation, in that many rounds of division may lie between the top and bottom of the cascade, and during these divisions, especially in the latter stages, progressive differentiation is evident. Hence the proerythroblast, erythroblast, reticulocyle, erythrocyte progression; this progression is both morphological, in that the cells come to resemble the erythrocyte end product more closely, and also biochemical in that products such as globin messenger RNA, globin and spectrin (a red cell surface protein) begin to appear.

Before we leave our study of the erythropoietic story, it seems appropriate to describe the experimental protocol which has provided most of the information summarised above. It involves exploitation of a remarkable technique developed by Lala and Johnson (1978). If mice are given a total body dose of high intensity X-rays, they die from loss of red and white blood cells, largely because the rapidly dividing erythropoietic tissue in liver and bone marrow is especially sensitive to irradiation. However, these mice can be rescued, by providing them with an injection of bone marrow cells donated by another unirradiated mouse of the same strain. If one of the rescued mice is examined by dissection some days after injection, the spleen will be found to have adopted a nodular appearance; when such nodules are themselves examined, each is found to consist of a colony of differentiating blood cells, some colonies consisting chiefly of granulocytic cells, others of erythroblasts, other cells of a lymphoid nature. By exploiting this fortuitous situation, especially by injecting discrete populations of erythroid stem cells, the nature of the erythropoietic cascade has been revealed, since spleen colonies can be chosen each of which represents a population resulting from division of one original stem cell or other cell further down the cascade. Thus, as will be deduced from Fig. 5.6, mixed colonies of granulocytes and macrophages can be found, but no colonies contain both macrophages and erythrocytes (or immediate erythrocyte precursor cells).

5.6 The stability of commitment is not absolute

In section 5.3, it was emphasised that cell commitment was normally very stable so that, once determination or differentiation is established, the cell and its progeny are highly resistant to further fate change. Now we must broaden our coverage to take into account the rare but highly significant exceptions to this rule. Two will be presented here, the first being an example of a change of commitment in determined cells, i.e. transdetermination, and the second a change of commitment in differentiated cells i.e. transdifferentiation. Although neither situation is entirely understood molecularly, it must be

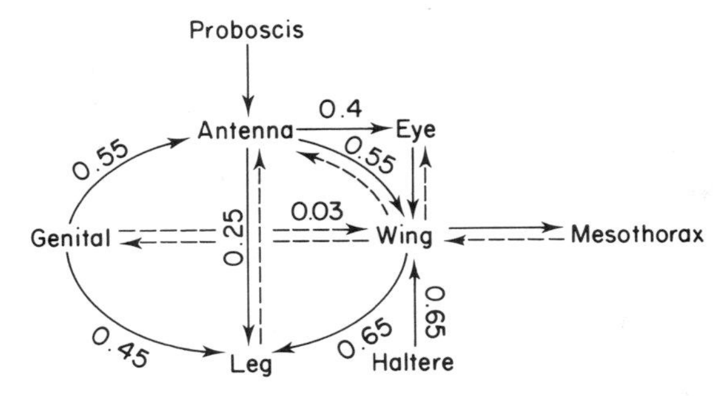

Fig. 5.7 The directions and frequencies of transdetermination in *Drosophila* imaginal discs. Those occurring most frequently are indicated by solid lines: those occurring less frequently are indicated by dashed lines. (Approximate frequencies are modified from Kauffman, S.A. (1973), *Science*, **181**, 310–18).

emphasised that neither is explicable by simple gene mutation. Indeed each involves the more or less synchronous change of a group of cells rather than one cell alone.

The transdetermination phenomenon is best known within the context of insect imaginal discs. This biological situation has already been cited as one displaying remarkable stability of commitment, yet at a low but significant frequency, a disc, or more often part of a disc, will terminate its dedication to one speciality and adopt another. Although the phenomenon does occur in Nature, it is most frequently encountered when discs are subcultured and cut up into separate pieces. What is particularly remarkable is that, when change does occur, it is not random. For example, a group of cells in a disc committed to be antenna, will now change to become committed to produce legs. As seen in Fig. 5.7 a probability matrix can be constructed reflecting the observed probabilities of transdetermination. Here then is an indication of some sort of regulatory machinary which, once in a while, can become involved in a switch, but not a switch to chaos or to any other genetic programme at random, but instead only to certain predetermined alternative genetic programmes.

Transdifferentiation is a more elusive phenomenon that transdetermination, largely because it is difficult to prove that the cells did not indulge in some sort of backtracking of commitment. But once more a particular and striking biological situation affords the prime example. It involves the regeneration of the lens of the eye following artificial lententomy (lens removal) in amphibians. Animals of this group have retained much greater general capacity for regeneration than higher vertebrates, and will readily regenerate tails and limbs, even as adults. But lens regeneration is remarkable because the new lens grows from previously pigmented cells of the iris. As illustrated in Fig. 5.8 what happens is that, following lens removal, epithelial cells at the edge of the iris, heavily pigmented with melanin, being to loose their somewhat stellar shape and conspicuous pigment. and instead begin to adopt a more elongated form. Eventually they acquire the linear form characteristic of lens fibre cells and proceed to

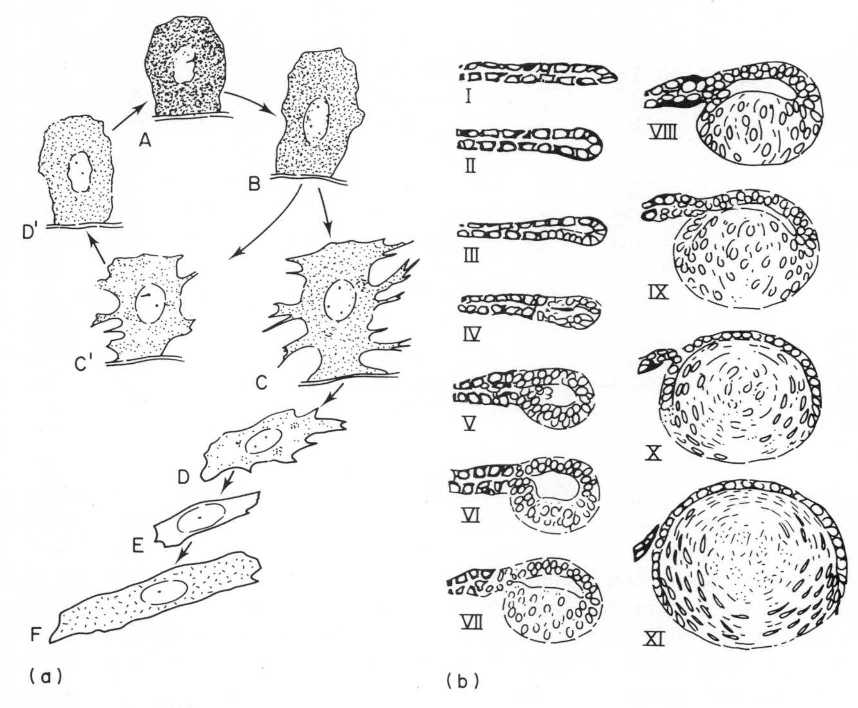

Fig. 5.8 Two pathways of development and differentiation in proliferating iris epithelial cells (ECs) of the newt. A-B-C-D-E-F, the pathway of conversion: A-B-C'-D'-A, the pathway of retrieval. Black dots, melanosomes: short lines, crystallins, (b) Lens regeneration in adult newts *N. viridescens*. The Roman numerals are Sato-stages (Sato. 1940). Each figure represents a median sagittal section of the eye. Only dorsal iris epithelium and regenerates are indicated. The pigmented cells are shown black. The surface alterations of IECs that occur during stages II-IX in pigmented IECs are not considered. (Reproduced, with permission, from Yamada, T. (1983), Control mechanisms in cell-type conversion in newt lens regeneration, in *Monographs in Developmental Biology*, **13**).

synthesise the protein which is uniquely characteristic of the lens, crystallin. Here then is a change of differentiated state with little or no retreat to an undifferentiated form. Presumably absence of the lens leads to signals being received by the iris cells which specifically induce the onset of this remarkable change. So both determination and differentiation. although in most cases remarkably stable, are not absolutely so.

5.7 The mechanism of commitment

Having reviewed some of the chief characteristics of cell commitment we are better equipped to return to consider how it is initiated. There seems little doubt that once it is established it is maintained by a balanced interaction and exchange of signals between nucleus and cytoplasm, which ensure that a predetermined set of genes is selectively expressed and others suppressed. But this does not explain how the gene set is chosen nor the choice of cell type made. Since the early embryo is the location in which many of the first signs of cell specialisation arise, it is appropriate to turn to embryology in our search for explanatory mechanisms. The truth turns out to be that there are many different mechanisms, and that those that are important in the commitment of one cell type are often different from those significant in another. The sort of mechanisms which are closely correlated with the onset of commitment are extrinsic signals such as gravity, that is whether a cell is at the top or bottom of an embryo, cell contact with other specialised cells, external ionic molecules, and other external environmental factors such as unequal cell divisions, and uneven gradients of molecules in a cell which does divide equally. At its simplest, all that one requires is some initial external signal and then to use existing inequalities to serve as further cues in commitment. This appears to be what happens in some organisms such as the small nematode worm *Caenorhabditis elegans* (see the remarkable cell lineage report by Sulston *et al.*, 1983), although even here within the lineage mapped out there is not absolute predetermination at all stages. However, in most situations a great number of cues are clearly utilised, and the phenomenon of commitment continues to occur even into adult life. What is particularly impressive is that cells are routinely being produced in different locations in the organism but specialised in precisely the same way. Thus studies with tetraparental mice (essentially a mouse which develops from an embryo which consists of a mixture of cells from two different embryos) demonstrate that many organisms have identical specialised cells which have derived from quite separate cells in the early embryo. So cells at all stages of development pass through times of decision when, in response to combinations of stimuli, decisions about destiny are made. These decisions often seem to be simple binary choices and, once made, they are interpreted and become manifest in terms of a set of expressed genes. No doubt there are gene regulatory mechanisms which help to stabilise the activity of specific selected sets of genes: the simplest of these mechanisms involves sequences which have come to be known as tissue master genes (see review by Maclean and Hilder, 1977).

5.8 The stabilisation of gene expression

Assuming that commitment is initiated as a response to signals that are primarily non-genetic, how are these signals translated into a fixed and remarkably stable programme of gene expression? The short answer must be that this still represents a grey area of molecular biology and remarkably little is known. But some things can be stated with reasonable certainty. The first is that in eukaryotes each gene involved in such differential orchestration seems to be regulated independently. The operons of prokaryotes are absent from eukaryotes. This independence is further emphasised by the finding that gene locus is rarely important in eukaryotic gene expression, so that genes involved in the same metabolic pathway are often located on different chromosomes, and when novel genes are artificially inserted into a genome by genetic engineering, their location rarely has implications for their activity or expression.

No tissue master genes of the type visualised in Maclean and Hilder (1977), and alluded to in the section above, have yet been identified for certain. But some of the homeo-box genes involved in *Drosophila* homeotic mutants may be candidates (see page 93) and so it is difficult to avoid the conclusion that the stability of a set programme of gene expression, so fundamental an aspect of cell differentiation, is a result of the activity of a few controlling genes. The trans-determination phenomenon in insect imaginal discs may well be an example of a transfer of control from one of these control sequences to another in each of the affected cells.

5.9 Some special mechanisms of gene regulation

In this and the preceeding chapter, the main mechanisms operative in eukaryotes are reviewed, and presented in the context of their probable involvement in cell commitment and differentiation. However, in certain specialised situations very particular and unusual processes are active in gene regulation, and some of these are too important to be overlooked. Only two situations will be discussed here. The first, gene amplification, affects a few particular genes in certain organisms, and often at very critical developmental stages, while the second situation involves a series of novel mechanisms all implicated in the production of immunoglobin molecules (antibody).

5.9.1 Gene amplification

Although not always used in this special confined sense, this terminology should best be reserved for situations in which particular gene sequences are temporarily increased in copy number as compared with other genes in the same genome. It has, to date, been discovered in three quite separate biological situations. They involve genes coding for chorion (egg coat) proteins in the *Drosophila* ovarian tissue, genes coding for metallothionein (metal binding proteins) in mice exposed to increased doses of certain heavy metals, and genes coding for the large ribosomal RNA molecules in some vertebrate oocytes. Only the latter situation will be described further.

The developing oocyte in an organism such as the amphibian *Xenopus* is an enormous cell responsible for building up stocks of molecules to be used by the early embryo after fertilisation. One molecular aggregate needed by the fertilised egg and early embryo is the ribosome, since only in association with ribosomes can new proteins be synthesised. So great is the demand for 28 and 18S ribosomal RNA in the oocyte that the 450 copies of these genes in the haploid genome are quite insufficient to meet the demand even at maximum transcription rates. To meet this increased requirement, the cells amplify the copy number of these genes preferentially so that between 1 and 2 million are present for a short developmental period in the oocyte. Essentially these exist within large numbers of mini-nucleoli, and enormously increase the output of large rRNA molecules. A number of points about this amplification are noteworthy. Firstly, it is temporary, the extra copies being rapidly degraded following maturation and fertilisation. Secondly, only the 28 and 18S ribosomal RNA genes are involved. The small 5S ribosomal genes, although they also make a necessary contribution to the ribosomes, are not specifically amplified. Rather, the copy number is set at a high level in the normal genome. And thirdly, gene amplification remains a rare phenomenon. Other genes whose products are in high demand in specialised cells, such as globin and fibroin genes in erythroid and silk gland cells respectively, have been clearly shown not to be amplified in these cells.

5.9.2 Immunoglobulin synthesis

Synthesis of the antibody proteins, the immunoglobulins, is rivalled only by the nervous system of higher vertebrates in terms of complexity. Many different types of lymphocytes are involved, and details of the immune system are quite beyond the scope of this small book. The appropriate chapter on the immune system in Alberts *et al.* (1983) is particularly good as a review of this area. What makes the immune system so complex in genetic terms is the requirement for an organism to have a potential for synthesis of an almost infinite number of different immunoglobulin molecules, each being elaborated by a distinct clone of cells within the lymphoid system. Three groups of special genetic mechanisms are involved, and we will consider each in turn.

(a) *Allelic exclusion*
Each lymphocyte which manufactures antibody makes a single species only, despite the fact that almost invariably the two allelic copies of the relevant genes will specify slightly variant forms. This feat is accomplished by the arrangement that in each cell only one allele for the appropriate immunoglobulin gene sequence is active. How this is achieved remains obscure.

(b) *V and C gene splicing*
In order to understand the next genetic curiosity of the immune system, some understanding of the structure of immunoglobulin (Ig) is necessary. As indicated in Fig. 5.9 the whole complex consists of four chains, two so-called light chains and two so-called heavy chains. Each of these four chains is made up of constant (C) region and a variable (V) region. It is the variable structure of these latter regions which account for Ig structural diversity. The amazing observation which underlies the juxtaposition of the V and C regions is that in non-lymphoid cells the sequences which specify these regions are not adjacent in the genome, yet they become contiguous in the relevant lymphoid cells (Tonegawa, 1985). What happens is that in the case of the light chain sequences, there is differential excision of

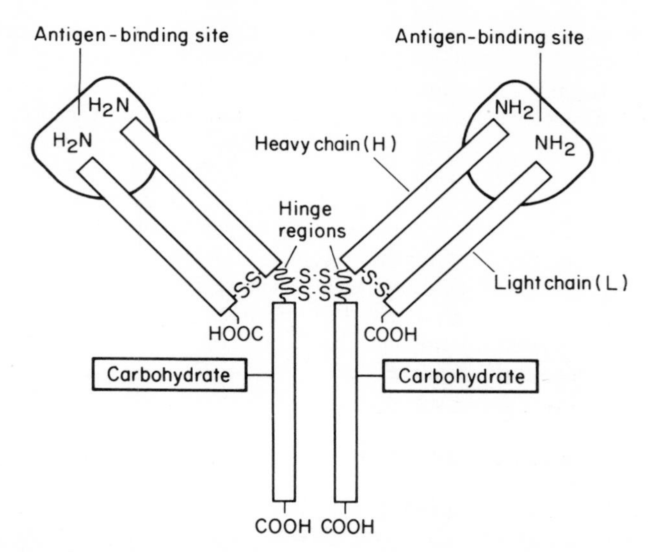

Fig. 5.9 An immunoglobulin molecule. The complex consists of two identical heavy chains (H) associated with two identical light chains (L). The antigen binding sites result from a complex of amino-terminal regions of both heavy and light chains. (Adapted, with permission, from Alberts, B. *et al. Molecular Biology of the Cell*, Garland, (1983)).

an interposing region and splicing of the relevant V and C regions in the appropriate lymphoid cell genome. In the case of the heavy chain sequences, fairly major transposition is required and this is also successfully engineered to permit read-through of the spliced sequences and generation of a single heavy chain transcript. So here we find some very precise gene manipulation functioning as a gene regulatory mechanism in cell differentiation.

(c) *Somatic mutation of V region genes*

The immune system has still more tricks in store for us. I mentioned in the paragraphs above that Ig molecules come in enormous variety, and that most of this variation is encoded in the structure of the V region sequence. It seems that even the large genomes of vertebrates do not carry a sufficient number of Ig coding sequences to generate the required diversity of Ig molecules (a mouse can probably make about 10 000 variant region Ig amino acid sequences by manipulation of the sequences present, whereas the number required has been estimated to be of the order of 10^8 or 10^9. How can this gap be closed? The answer is by preferential somatic mutation. There is now evidence that a high rate of somatic mutation in the V region sequences is exploited to increase their diversity in different cells. Thus the cells of the immune system have evolved special and probably unique genetic tricks to accomplish the enormous demands of antibody structure and production. To date none of these mechanisms is known to be utilised routinely by any other set of gene sequences.

6

Genes in the Context of Evolution

6.1 Evolution at a genetic level

In this short concluding chapter I will endeavour to develop a view of genes and gene regulation within the framework of evolution. It seems to me important to do so, partly because genes have become what they are as a result of responses to evolutionary pressures, and also because it is as a result of initial changes in the genome that the phenotype comes to provide the material on which the evolutionary selection acts. In other words, selection is at the level of the phenotype, primarily, but mutations, which provide the changes on which the forces of evolution act, occur in the DNA. Firstly, let us look at the mechanics of mutation, how it occurs, and how the organism responds to it.

It is important to differentiate between damage to DNA and mutation, since only some such damage results in mutation. This is partly because some damage is quickly repaired by specific repair enzymes and so does not lead to permanent mutation, and also because some damage is rapidly lethal to the affected cell and so does not induce a persistent mutation.

6.2 Mutation is a random process

Mutation is a rare event when considered in the context of a single base in the DNA or even a single gene, but is a common event in terms of the whole genome, or, even more obviously, in the context of the gene pool of a whole population of organisms. It is crucial to understand at the outset the triple aspects of its random nature. Mutation is random in that it constitutes an exception to the normal regularity of DNA structure and replication, in that it cannot be forecast in terms of a particular gene, organism or generation, and that it is not adaptive. This last point is particularly noteworthy, and implies that although natural selection favours advantageous mutations and operates against deleterious ones, there is no mechanism whereby an organism can produce a beneficial mutation by demand, as it were.

Mutation rates may be expressed in a number of different ways, depending on standpoint and usefulness. Therefore, it can be demonstrated that, in general terms, one base change occurs in each round of DNA replication per

10^9 base pairs. As a function of scoring the number of mutations likely to occur in any one gene in a single organism per generation the figure varies from 1–3 $\times$ 10^{-9} for *E. coli* to between 1 and 10 $\times$ 10^{-6} for the human and many other eukaryotes – the reasons for the disparity between bacteria and eukaryotes and the range within each are complex, (see discussion in Ayala and Kiger, 1980). However, a few relevant points should be emphasised. Firstly, that mutations in DNA are of many kinds, including loss or addition of bases, or substitution of one base by another (the most common). If we enquire how such mutations occur then we find that various chemicals, such as nitrogen mustard, are potent mutagens, as are physical factors such as irradiation by X-rays, cosmic rays or other radio emissions. But as we have observed, the relationship between DNA damage and mutation rate is complex, because much of the damage encountered by the DNA is repaired by specific DNA repair enzymes (see Section 6.3 below). Mutations may occur in germ cells or somatic cells, but only the former will be detectable in the progeny (and then only when the mutation is dominant over wild type or appears in a homozygous condition). Indeed somatic mutation is particularly hard to measure since the particular gene affected is unlikely to be expressed in the cell in which the mutational event has occurred. It is likely that somatic mutation is often important in the induction of cancer (see discussion of oncogenes at the end of this chapter on page 126).

Mutation can also be usefully subdivided into changes at a chromosomal level and those occurring at the level of DNA and affecting only one or a few bases. Chromosomal mutations may involve deletion, duplication, or inversion of a chromosome segment, or translocation of a chromosome segment from one chromosome to another, or a change in the numbers of whole chromosomes. We will not dwell further here on the topic of chromosome mutations and the interested reader should consult Chapter 17 in the excellent textbook by Ayala and Kiger (1980).

6.3 DNA repair ensures that most damage to DNA does not result in mutation

In 1945, the physicist Erwin Schroedinger pointed out that genes must be prone to constant damage through thermal collisions with solvent molecules in the cell. This observation preceded the discovery of DNA as the genetic material but still holds true. For example, it has been estimated that a single human cell loses approximately 5000 purine bases per day, while ultraviolet light causes the rapid development of thymine dimers within the DNA. It therefore becomes clear that if cells did not have rapidly evolving mechanisms of DNA repair, they would have immediately succumbed to the intrinsic instability of DNA within the milieu of the cell. DNA repair processes are so efficient that they reduce the level of damage to a tiny fraction of its original level, and this is 'allowed to persist' as mutation. The phrase, 'allowed to persist' is used advisedly, since there is good evidence that cells are capable of repairing DNA more efficiently than they do. For example, one of the DNA repair mechanisms is said to be error-prone, and the level of error-proneness effectively sets the mutation rate. This probably explains in large measure why mutation rates are so variable between different groups of organisms. For example, one bacterial species, *Mycobacterium radiodurans*, is known to be able to tolerate a level of radiation many times the human lethal dose, not because it tolerates mutation, but because it has hyperefficient DNA repair mechanisms. An example of the serious consequences of a failure of DNA repair mechanisms is provided by the human disease xeroderma pigmentosa, in which affected individuals develop large numbers of skin cancers. These result from non-repaired lesions in the DNA following exposure to ultraviolet irradiation.

Since mutation has to be understood within the context of an error-prone repair process, it will also follow that mutation rate is itself open to evolutionary selection. Since mutation provides the basis for the variation on which natural selection works, it will become clear that hyperefficient repair enzymes will tend to suppress both mutation and variation, and therefore the basis for natural selection. This might be of immediate benefit to the individual but be irreconcilable with the future survival of the population. On the other hand, ineffective repair would tend to allow mutation rates to be high, with consequent high variability but increased chances of the appearance of numerous deleterious mutations. So organisms not only evolve, but can, in the long terms, control their own rate of evolution.

A number of distinct types of DNA repair occur in cells, and these are now briefly outlined. All of these repair processes depend on the fact that DNA is normally double-stranded, and the damage is recognised and repaired with reference to the undamaged strand. Only an outline of DNA repair can be attempted here, since it has been estimated that, even in a primitive eukaryote such as yeast, at least 50 different repair enzymes are implicated in DNA repair.

a) *Removal of 'bulky lesions'* These DNA lesions include the thymine dimers introduced by sunlight and UV light, and covalent reactions of DNA with molecules such as hydrocarbon carcinogens. In both of these cases the damage leads to a major distortion or bulge in the double helix, and this is recognised by a multienzyme complex. The DNA backbone is cut on either side and around the distortion and the lesion excised by enzyme action. DNA polymerases then make a new good copy by base pairing with the existing undistorted strand, followed by sealing of the nicks by DNA ligase to complete the repair.

b) *Repair of depurination* This is by far the most common form of damage and repair to DNA. Repair nucleases recognise that a base is missing, and proceed to remove other adjacent bases in the same strand, along with the appropriate pieces of phosphodiester backbone. Once more an intact and correct sequence is inserted by polymerase enzymes as discussed for 'bulky lesion' repair, and ligase enzymes, effect the ligation of the new inserts resulting in a colinear strand.

c) *DNA glycosylase repair* DNA glycosylases are a group of over twenty distinct enzymes that recognise defective bases such as deaminated adenines or cytosines, alkylated bases, or bases with defective carbon bonds. Each enzyme catalyzes the hydrolytic removal of the defective base, cleaving it away from the deoxyribose sugar in the backbone. Nuclease enzymes then recognise the missing base, and cut out the relevant piece of backbone plus neighbouring sections, and this is followed, as in (a) and (b), by polymerase catalyzed addition of new nucleotides and ligation by ligase.

6.4 The rate of change in the DNA may be quite different from that in the proteins coded by the DNA

When it was realised that DNA was the genetic material of all living organisms, scientists began to compare organisms in terms of their composition of the four bases in DNA. To their initial surprise, even closely related organisms with few amino acid differences in any of their proteins often revealed quite marked divergence in base composition. We now realise that this results from an interplay of a number of different factors. Firstly, much of the DNA of the higher eukaryotes is not coding sequence, and there is probably little or no selective pressure to prevent the rapid accumulation of base changes as a result of mutation. This is exemplified by comparing the coding sequences for the large ribosomal RNAs in two species of *Xenopus*, namely *laevis* and *borealis*. In these two species these sequences are strikingly similar. But if the spacer sequences between these coding sequences are compared, dramatic differences are apparent, (Moss et al. 1985).

The second factor has already been discussed in Chapter 1, namely the degeneracy of the code, especially of the third letter of the anticodons in the DNA sense strand. So it is quite possible for a gene sequence to accumulate a substantial number of base changes with no change in the amino acid sequence of the protein whatever. If these two factors are taken together it will be clear why the overall base composition of DNA of closely related species may often be quite divergent, and this divergence will also be apparent if generalised DNA sequence is compared. That is why the randomised comparison of DNA sequence has proved so disappointing to taxonomists eager to find in the DNA an accurate indicator of the extent of the evolutionary relationship between two organisms. Yet some of the changes in the DNA must lead to changes in amino acid sequence, otherwise no evolution at a phenotypic level would occur. We must therefore now examine how such mutation occur and become fixed in the gene pool of a population.

6.5 The relationship between gene mutation and protein structure

One of the first aspects of mutation that becomes evident when comparisons are made between the structures of analogous proteins from different organisms is the great divergence in the rate of accumulated change between different proteins. For example, the core histones H3 and H4 have scarcely changed at

name	residue		replacement		location in hemoglobin chain or tetramer	abnormal property	clinical effect in the heterozygote
	number	position	from	to			
J Toronto	5	A3	Ala→Asp		External	–	–
J Paris	12	A10	Ala→Asp		External	–	–
J Oxford	15	A13	Gly→Asp		External	–	–
I	16	A14	Lys→Glu		External	–	–
J Meddellin	22	B3	Gly→Asp		External	–	–
Memphis	23	B4	Glu→Gln		External	–	–
G Audhali	23	B4	Glu→Val		External	–	–
Chad	23	B4	Glu→Lys		External	–	–
G Fort Worth	27	B8	Glu→Gly			–	–
G Chinese	30	B11	Glu→Gin		α, β^1 contact		
Torino	43	CD1	Phe→Val		Heme contact	Instability	Hemolytic anemia
L Ferrara	47	CD5	Asp→Gly		External	Instability	Mild hemolytic anemia
Sealy-Sinai-Hasharon	47	CD5	Asp→His		External	Instability	Mild hemolytic anemia
J Sardegna	50	CD8	His→Asp		External	–	–
Russ	51	CD9	Gly→Arg		External	–	–
Shimonoseki	54	E3	Gln→Arg		External	–	–
Mexico	54	E3	Gln→Glu		External	–	–
L Persian Gulf	57	E6	Gly→Arg		External	–	–
Norfolk	57	E6	Gly→Asp		External	–	–
M Boston	58	E7	His→Tyr		Heme contact distal distidine	Methemoglobin	Cyanosis
Zambia	60	E9	Lys→Asn			–	–
Buda	61	E10	Lys→Asn			–	–
G Philadelphia	68	E17	Asn→Lys		External	–	–
Ube-2	68	E17	Asn→Asp		External	–	–
G Taichung	74	EF3	Asp→His			–	–
Q Iran	75	EF4	Asp→His			–	–
Stanleyville-2	78	EF7	Asn→Lys		External	–	–
Ann Arbor	80	F1	Leu→Arg		Internal	Instability	Hemolytic anemia
Etobicoke	84	F5	Ser→Arg		External	–	–
M Iwate	87	F8	His→Tyr		Heme contact proximal histidine	Methemoglobin	Cyanosis
Broussals	90	FG2	Lys→Asn		External	–	–
J Rajappen	90	FG2	Lys→Thr		External	–	–
J Capetown	92	FG4	Arg→Gln		$\alpha_1\beta_2$ contact	↑ O_2 affinity	
Chesapeake	92	FG4	Arg→Leu		$\alpha_1\beta_2$ contact	↑ O_2 affinity	Polycythemia
G Georgia	95	G2	Pro→Leu		$\alpha_1\beta_2$ contact		
Rampa	95	G2	Pro→Ser		$\alpha_1\beta_2$ contact		
Manitoba	102	G9	Ser→Arg		External	–	–
Chiapas	114	GH2	Pro→Arg			–	–
J Tonganiki	115	GH3	Ala→Asp		External	–	–
O Indonesia	116	GH4	Glu→Lys		External	–	–
Bibba	136	H19	Leu→Pro		Heme contact	Instability	Severe hemolytic anemia
Singapore	141	HC3	Arg→Pro		$\alpha_1\alpha_2$ contact		

Symbols: ↑ – increased; ↓ – decreased: – – no abnormality

Fig. 6.1 A series of abnormal human haemoglobins which have resulted from amino acid substitutions in the alpha globin chain. (Reproduced with permission from Wagner, R.P. *et al.*, *Introduction to Modern Genetics*, Wiley, (1980)).

all in the course of evolution, while others, such as the globin moiety of haemoglobin have changed considerably. The reason lies in the relationship between protein structure and function. If the entire length of a polypeptide chain is closely involved in protein function, say in forming a ligand binding site, then probably even minor base changes which would alter one amino acid might prove irreconcilable with the activity of the protein. So it is with the core histones, which are involved in binding to one another, to other histones within the nucleosome, and to the DNA.

On the other hand, in proteins such as haemoglobin the haem moeity is bound in a special pocket of the folded globin chain. Therefore one might expect that the globin molecule could readily tolerate amino acid substitutions in the non-pocket area, while being rather resistant to them in the critical haem-binding pocket. And, as seen in Fig. 6.1, this is exactly what we do find. So, as a generalisation, active sites of proteins are rather intolerant of change, other regions may well diverge quite quickly.

It must also be borne in mind that some amino acid substitutions are relatively conservative, so that one amino acid is replaced by another with very similar properties. This is what we find when we scrutinise the changes that have occurred in the course of evolution to the histone H4. In the histone H4 of pea, the 60th amino acid is isoleucine and the 77th is arginine. These are replaced in the bovine molecule by valine and lysine respectively: both isoleucine and valine are apolar residues with similar hydrocarbon side chains, while both arginine and lysine are basic residues. Such conservative substitutions as a result of mutation in the DNA are likely to result in minimal changes in the properties of the resulting protein. Other non-conservative substitutions, as for example, substitution of an arginine (basic) for a glutamic acid (acidic), could have much more profound effects on proein structure and function.

A third point to note is that diverence of gene sequences and of the proteins which they specify is much easier if more than one copy of the gene is present in the genome. Therefore gene duplication is often a prerequisite to genetic evolution. So it is with the globin genes, where a small family of sequences have now evolved in mammals, undoubtedly from the original duplication of an ancestral globin gene. This has afforded mammals the opportunity to exploit the slightly differing capacities of the variant globins, providing opportunity for the development of distinct embryonic, foetal and adult haemoglobins, through utilising particular combinations of the available globins (see Fig. 6.2). Remember, of course, that most mutations will be deleterious just on the basis of the random substitution of one amino acid to another. So organisms with deleterious changes expressed in the phenotype wll be at a slight disadvantage in the population and will tend to die out from the population with time, while the rare organisms with beneficial phenotypic changes will be positively selected and will gradually spread their genes through the population gene pool.

Much argument had occurred in the literature over the possible occurrence of so-called neutral mutations, that is mutations that lead to phenotypic changes that are neither beneficial nor deleterious. This view has been strongly

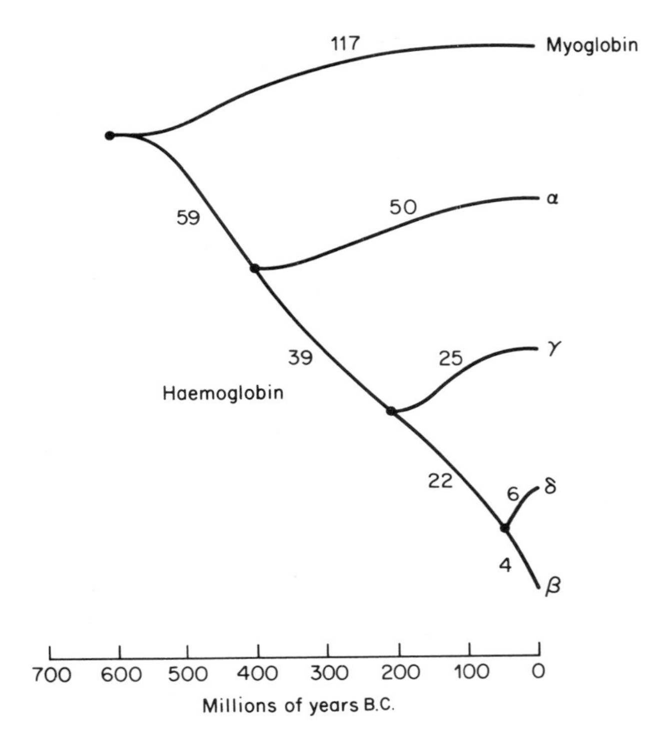

Fig. 6.2 Suggested evolutionary history of the globin genes from a single ancestral gene. Dots indicate where gene duplication led to the appearance of two distinct globin gene copies. Numbers indicate the minimum number of nucleotide substitutions required to account for observed amino acid differences between branches. The first duplication is presumed to have occurred some 600×10^6 years ago – based on comparative morphology and paleantology. (Reproduced with permission from Ayala, F.J. and Kiger, J.A., *Modern Genetics*, second edition, Benjamin, (1984)).

advocated by Kimura (see Kimura and Ohta (1971)) and will not be discussed further here.

The number of amino acid differences in the structure of an individual protein can be used to indicate a phylogenetic relationship, and in many cases, for example, cytochrome c (see Fig. 6.3), this agrees well with the evolutionary relationships between organisms that are extrapolated from the fossil record. Of these differences presumably none are actually deleterious, and at least some are advantageous, the remainder being either neutral or slightly advantageous depending on how one interprets the arguements for 'neutralism' – I personally find them persuasive. It must also be borne in mind that relatively few of the possible changes in polypeptide chain sequence are likely to prove advantageous or even neutral. Since each amino acid can be replaced by another 19 amino acids, each chemically distinct and with individual properties, the total

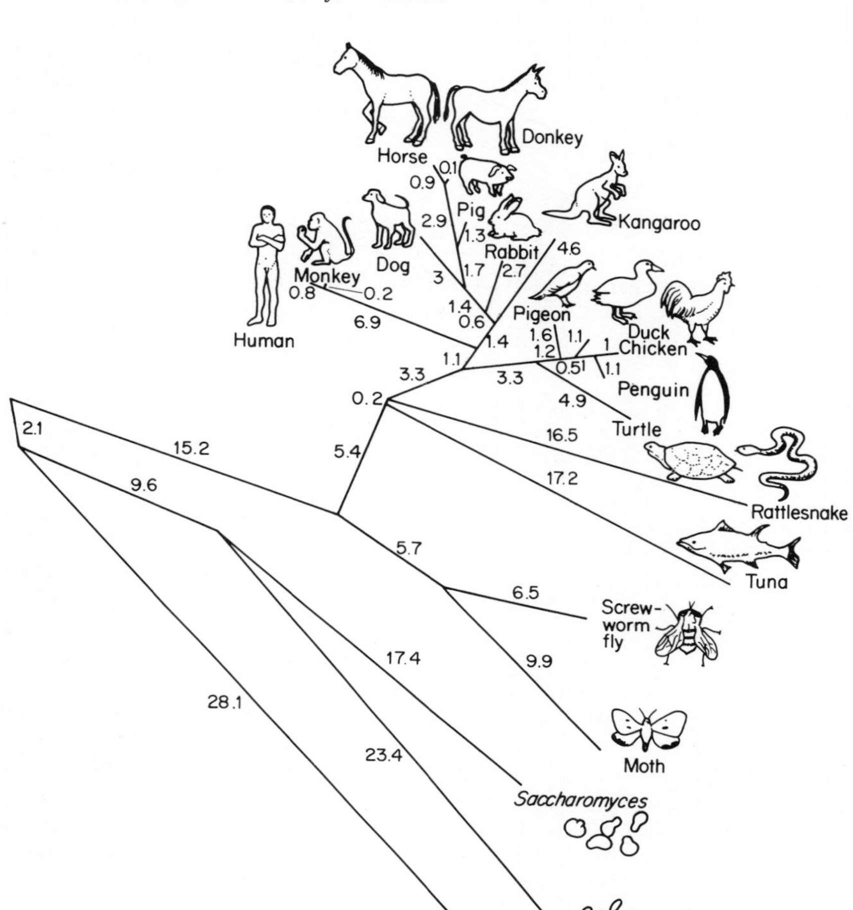

Fig. 6.3 Suggested phylogeny of 20 organisms, based on observed differences in the amino acid sequence of cytochrome C. The minimum number of nucleotide changes required for each branch is indicated. The tree is in rough agreement with the evolutionary relationship data derived from the fossil record. (Reproduced with permission from Ayala, F.J. and Kiger, J.A., *Modern Genetics*, second edition, Benjamin (1984)).

number of variants is enormous. For example, a molecule with 170 amino acids, such as the globins, 2×10^{170} possible variant structures could theoretically be generated. Of these, taking all the vertebrates together, there are possibly only a few hundred variant structures in the globin family, and, because of the limitations mentioned for active sites and ligand binding, much of the existing variation is confined to particular parts of the polypeptide chain. The conservative nature of proteins follows from the fact that, in almost all cases, their structure and function are both dependent on elaborate three dimensional folding and the eventual structure must be sufficiently stable to carry out its cellular functions, with stability and precision, often over a lengthy period of time. From this we can deduce that only a tiny fraction of the mutations that escape repair are ever likely to survive for long in a population of organisms, at least as far as the structural gene sequences are concerned. Probably the rates of change in some other sequences are very much higher.

Before we leave the topic of the relationships between genetic evolution and protein structure, it should be stressed that new proteins can readily be generated

(i) by simply aggregating a number of polypeptide chains to form a large multi-subunit structure (as has happened, for example, with haemoglobin and bacterial RNA polymerase) and

(ii) by new assortments of existing protein domains. Most eukaryotic genes are split, and it is likely that many entire exon sequences code for specific protein domains, each of which has acquired, through evolutionary selection, a high degree of independent efficiency as, say, a specific ligand for a particular molecule. Evolution could then proceed to derive new proteins with new combinations of function by shuffling existing exons into unique new combinations and splicing them together with intron sequences. There is good evidence that many proteins do indeed share existing domains with other dissimilar proteins in the same organism. Not only does this permit the association of two previously separate and distinct binding sites, but the region of association between two such domains in the new structure may produce an absolutely novel and useful binding site. So a new protein made by a novel combination of existing domains might constitute a startling advance in some aspect of cellular metabolism.

6.6 Selfish DNA

In his most readable and provocative book *The Selfish Gene* (1976), Richard Dawkins has enlarged on an idea also discussed by Orgel and Crick (1980), namely that much of the DNA is eukaryotic genomes may have no genetic function and be of no use to the organism. In these senses it has been dubbed 'selfish DNA' by these authors, since it can be seen to utilise the genome and the organism within which it is located as a replicating system, thus ensuring its own survival and proliferation.

In my view we do not know enough about the molecular structure and function of this material to be sure that this proposition is correct, but it is certainly a stimulating hypothesis. A look at Table 1.1 will remind us that many higher eukaryotes have extravagantly large genomes, and as was emphasised in Chapter 1, only a small proportion of this DNA can serve as either structural gene or regulatory sequence, at least in the strict interpretation of these terms.

This DNA is not satellite DNA, most of which is made up of very short sequences and concentrated around the centromeric regions of the chromosomes; instead this DNA, which comprises about 20% of mammalian DNA (and much more in some species) is interspersed throughout the genome, and is made up for the most part of sequences of some 300 nucleotides long. Some of these sequences are known to be transcribed, both as short transcripts and as parts of the longer hnRNA transcripts, but almost all of the RNA is purely nuclear in location. Many of these sequences are also known to function as transposons, (see Glossary) which may help to explain their widespread distribution in the genomes of higher organisms. It has therefore been suggested that much of this DNA can be viewed as being 'selfish', spreading through eukaryotic genomes because of its ease of transposition but serving no useful function in the genome. Many such sequences may function like introns, and thus their transposition into a coding sequence would not jeopardise its function and their presence would not constitute a selective disadvantage. Whether these multiple repeat sequences are indeed an example of selfish DNA remains uncertain, but interestingly they have in recent years gained a special significance in providing the basis for a new and powerful technique in molecular genetics – the process of DNA fingerprinting.

6.7 DNA fingerprinting provides a way to determine relationships between individuals very precisely

Many of the repeat copies of the interspersed sequences often described as selfish DNA exist in the genome as series of tandem repeats, the precise number of the series in a particular repeat being rather variable. The reason for this is presumed to be the frequency of unequal crossing over at meiotic recombination. Imagine a pair of homologous chromosomes each carrying, in a small section of one chromosome, a row of six repeat copies in tandem array. Since some parts of these repeats are closely conserved and so possess identical sequences, crossing over between such identical sequences, say in the 2nd copy of one chromosome and the 4th copy of the other, would result in one chromosome having four copies and the other eight. It seems that this may be a frequent occurrence at meiosis in so far as these interspersed tandem repeats are concerned. It also means that a particular parent is likely to differ from the other parent in terms of the numbers of these repeats, while the children would be expected to show patterns particularly characteristic of both parents. Dr. Alex Jefferys of Leicester University has developed this observation into a technique called DNA fingerprinting. DNA from the individual to be examined is purified and digested with a restriction endonuclease enzyme which cuts only outside the repeated sequence. The DNA fragments which result from this digestion are then separated on an electrophoretic gel. A radioactive probe DNA is then made using cloned sequences of the internal conserved sequence of the repeats, and the gel hybridised with this probe DNA prior to the production of an autoradiograph. As seen in Fig. 6.4, development of such an autoradiograph reveals hybridisation of the probe with bands of DNA of a great

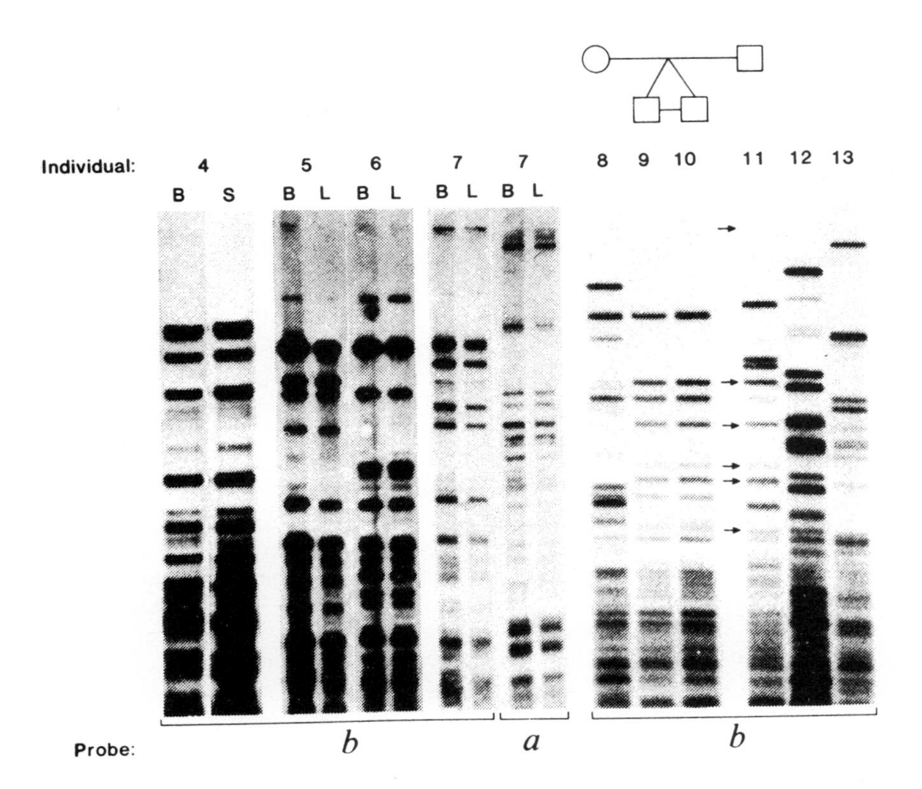

Fig. 6.4 *DNA fingerprints from a series of human individuals*. DNA samples have been digested with appropriate restriction enzymes, electrophoresed through agar gels, and hybridised with radioactive probe DNA from a repeated sequence of human hypervariable satellite DNA region. Bands therefore represent DNA from such satellite regions which show homology with the probe. DNA from 13 different human individuals is shown, as indicated by numbers above the lanes, while the letters above the lanes indicate where the DNA was obtained from blood (B), sperm (S) or white cells (L). All the 6 right hand lanes are from blood. The letters below the lanes indicate that two separate probes were used, probe (a) for only two samples, and probe (b) for all others.

Notice that the band pattern is essentially identical when the same probe is used with DNA derived from different tissues but the same individual (as in first two lanes), but that different probes give quite distinct patterns (compare first pair and second pair from sample 7) even with DNA from the same individual.

Of greatest interest is that 9 and 10 are derived from two identical twins (and are essentially identical in fingerprint pattern), and should be compared with lane 8 (their mother) and lane 11 (their father), with both of whom they have bands in common. 12 and 13 are from two unrelated men and have no common bands with 8, 9, 10 or 11. Similarly but less clearly, 6 is the daugther of 5 and 7 the maternal aunt of 5 (note some common bands between 5 and 7) while 4 is unrelated.

Reproduced with permission from Jeffrey's *et al*., Nature 316 p. 78, 1985.

range of sizes, and analysis of such patterns reveals close identity between siblings, parents and progeny, and much less lesser identity between parents themselves. The technique has proved useful in providing evidence in cases of disputed paternity and also in determining relationships between individuals within wild populations of species such as birds and mammals. See the references under Jeffreys et al. (1985a, 1985b) and Burke and Bruford (1987).

So although analysis of the DNA in the genome is not a good way to determine evolutionary relationships between organisms in general, analysis of particular sequences can be used for determining very close relationships. It is also noteworthy that in recent years some success has been achieved in the use of restriction enzyme digests of mitochondrial DNA as a means of establishing relationships between populations and species (Skibinski, 1987).

6.8 The evolution of oncogenes

In Chapter 2, I mentioned that some, but not all, retroviruses contained an oncogene sequence, and that these viral oncogenes bear a striking resemblance to certain cellular genes, often termed protooncogenes. These latter sequences appear to be perfectly normal cellular genes, coding for proteins such as kinase enzymes or growth factors, and it is now widely supposed that when a retrovirus infects a cell, some disturbance of gene regulation may affect the expression of the appropriate protooncogene, the malignant state being a presumptive result of such disturbance. Retroviruses have two interesting implications for gene evolution. Firstly, they have presumably acquired the oncogene sequences by acquisition of an originally intact cellular gene, now dubbed a protooncogene. Secondly, as the unique possessors of reverse transcriptase enzyme, which makes DNA from an RNA template, their presence in cells makes possible the production of DNA sequences by reverse transcription of message. The processed pseudogenes seem likely candidates for sequences which have been incorporated into the genome by such a process, and presumably this must occur in germ line cells to ensure their persistance.

References

Alberts B., Bray, D., Lewis, J., Raff, M., Roberts K. and Watson, J.D., (eds) (1983). *Molecular Biology of the Cell.* Garland, New York.

Allfrey, V.G. (1977) Post synthetic modifications of histone structure. In *Chromatin and Chromosome Structure,* 167–91 H.J. Li and R. Echhardet (eds). Academic Press, New York.

Ayala, F.J. and Kiger, J.A. (1980) *Modern Genetics.* Benjamin/Cummings Publishing Company Inc. USA.

Bird, A.P. (1983) DNA modification in *Eukaryotic Genes: Structure, Activity and Regulation* (eds) N. Maclean, S.P. Gregory and R.A. Flavell. Butterworths, London.

Bogenhagen, D.F., Sakonju S. and Brown D.D. (1980) A control region in the centre of 5S RNA gene direct specific initiation of transcription, *Cell,* **19**, 27–35.

Bradbury, E.M., Maclean, N. and Mathews, H.R. (1981) *DNA, Chromatin and Chromosomes.* Blackwells, Oxford.

Burke, T. and Bruford, M.W. (1987) *DNA finger-printing in birds. Nature,* **327**, 149–152.

Butler, P.J.G. and Thomas, J.O. (1980) Changes in chromatin folding in solution. Journal of Molecular Biology, **140**, 505–529.

Compton, J.L. and McCarthy, B.J. (1978) Induction of the *Drosophila* heat shock response in isolated polytene nuclei. *Cell,* **14**, 191–202.

Dawkins, R. (1976) *The Selfish Gene.* Oxford University Press, Oxford.

Elgin, S.C.R. (1981) DNase I hypersensitive sites of chromatin. *Cell,* **27**, 413–415.

Etkin, L.D., and Diberardino, M.A. (1983) Expression of nuclei and purified genes microinjected into oocytes and eggs. In *Eukaryotic Genes: Structure, activity and regulation,* 127–56, (eds) N. Maclean, S.P. Gregory and R. Flavell. Butterworths, London.

Flavell, R.A. and Grosveld, F.G. (1983) Globin genes: their structure and expression in *Eukaryotic genes: structure, activity and regulation.* (eds) N. Maclean, S.P. Gregory and R.A. Flavell. Butterworths, London.

Fox, T.D. (1984) Multiple forms of mitochondrial DNA in higher plants. *Nature,* **307**, 415.

Gurdon, J.B. and Melton, D.A. (1981) Gene transfer in amphibian eggs and oocytes. *Annual Review of Genetics,* **15**, 189–218.

Hammerling, J. (1963) Nucleo-cytoplasmic interactions in *Acetabularia* and other cells. *Annual Review of Plant Physiology,* **14**, 65–92.

Harris, H. (1970) *Cell Fusion.* Clarendon Press, Oxford.

Jacob, F. and Monod, J. (1961) Genetic regulatory mechanics in the synthesis of proteins. *Journal of Molecular Biology,* **3**, 318–356.

Jeffreys, A.J., Wilson, V., and Thein, S.L. (1985a) Hypervariable 'minisatellite' regions in human DNA. *Nature*, **314**, 67–73.

Jeffreys, A.J., Wilson, V., and Thein, S.L. (1985b) Individual-specific 'fingerprints' of human DNA. *Nature*, **316**, 76–79.

Kimura, M. and Ohta, T. (1971) *Theoretical Aspects of Population Genetics*. Princeton University Press.

Lala, P.K. and Johnstone, G.R. (1978). Monoclonal origin of B-lymphocyte colony-forming cells in spleen colonies formed by multipotential haemopoietic stem cells. *Journal of Experimental Medicine*, **148**, 1468–77.

Levinger, L. and Varshavsky, A. (1982) Selective arrangement of ubiquitinated and DI protein - containing nucleosomes within the Drosophila genome. *Cell*, **28**, 375–385.

Maclean, N. and Hilder, V.A. (1977) Mechanisms of chromatin activation and repression. *International Review of Cytology*, **48**, 1–54.

Maclean, N., Gregory, S.P. and Flavell, R.A. (1983) *Eukaryotic Genes: Structure, Activity and Regulation*. Butterworths, London.

Maclean, N. and Hall, B.K. (1987) *Cell commitment and differentiation* Cambridge University Press. Cambridge.

Maxson, R., Mohun, T. and Kedes, L. (1983) 'Histone genes' in *Eukaryotic Genes: Structure, Activity and Regulation* (eds) Maclean, N., Gregory, S.P. and Flavell, R.A. Butterworths, London.

Miller, J.R. (1983) 5S Ribosomal RNA genes in *Eukaryotic Genes: Structure, Activity and Regulation*. (eds) N. Maclean, S.P. Gregory and R.A. Flavell. Butterworths, London.

Moss, T., Mitchelson, K. and Dewinter, R. (1985) The promotion of ribosomal transcription in eukaryotes. *Oxford Surveys on Eukaryotic Genes*, **2**, 205–250.

Orgel, L.E. and Crick, F.H.C (1980) Selfish DNA: the ultimate parasite. *Nature*, **284**, 604–607.

Picard, D. (1985) Viral and cellular transcription enhancers. *Oxford Surveys on Eukaryotic Genes*, **2**, 24–48.

Skibinski, D.O.F. (1985) Mytochondrial DNA variation in *mytilus edulis L.* and padstow mussel, *Journal of Experimental Marine Biology and Ecology*, **92**, 251–258.

Sulston, J.E., Schievenberg, E., White, J.G. and Thomson, J.N. (1983) The embryonic cell lineage of the nematode *Caenorhabditis elegans*. *Development Biology*, **100**, 64–119.

Tonegawa, S. (1985) The molecules of the immune system. *Scientific American*, **253**, 104–13.

Weintraub, H. and Groudine, M. (1976) Chromosomal subunits in active genes have an altered conformation. *Proceedings of the National Academy of Science*, **74**, 4867–4871.

Glossary

Allele (allelomorph) Any one of the varied forms of a gene found at the same locus on homologous chromosomes. Alleles may differ by one or many bases.

Autoradiography A technique in which a photographic image, the autoradiograph, is derived from localized radioactive molecules. These molecules may indicate a particular localization on an electrophoretic gel, or a specific site of localization in a histological specimen.

Autosome Any chromosome other than the sex chromosome (X and Y)

Cell cycle The life history of a cell between mitotic divisions, consisting of the first growth phase (G1), a DNA replication phase (S), a second growth phase (G2) and mitosis (M).

Centromere The part of a chromosome involved in spindle attachment. The DNA in the centromeric region is highly repetitious.

Chromatid One of the two halves of a chromosome resulting from DNA replication during the S phase of the cell cycle. They first become visible at mitotic or meiotic prophase.

Chromatin The material of which chromosomes are composed, made up chiefly of equal proportions of DNA and the basic protein, histone, together with lesser amounts of RNA and other proteins.

Chromomere Granules of condensed chromatin visible in prophase meiotic chromosomes. The dark bands on polytene chromosomes are chromomeres.

Cistron A gene which codes for a polypeptide chain. The word derives from the definition of a gene in terms of the cis-trans test.

Complementary DNA (cDNA) Coding sequence made by reverse transcription from messenger RNA. cDNA is therefore similar to the genomic gene sequence exons, but totally lacks intron sequences.

Conjugation A phenomenon in bacteria in which genetic material is passed from one bacterium to another.

Denaturation Loss of tertiary structure. Protein denaturation is usually irreversible but DNA may be dissociated into single strands and then renatured again by the reformation of hydrogen bonds between complementary bases.

Domain A functionally or structurally distinct part of a protein.

Downstream In the direction of DNA transcription. Downstream sequences are at the three prime end of the message or below, since the messenger RNA is transcribed from the five prime to the three prime end. Termination codons lie downstream with respect to the gene sequence which they terminate.

Enhancer A specific sequence that enhances the efficiency of transcription of a gene, often in a tissue specific manner. Enhancer sequences may lie upstream or downstream from the gene coding sequences.

Erythropoietic Relating to blood cell production.

Eukaryote An organism with nucleated cells. These include animals, plants and fungi, but exclude bacteria which are classed as prokaryotes.

Exon Part of a gene which is represented in the messenger RNA and is therefore translated into protein. In many genes exons are separated by introns, which are not represented in the message, being specifically cleaved out of the hnRNA.

Foldback DNA DNA which is able to form a partly or entirely base paired hair pin-like structure by folding back on itself when single stranded. Such DNA renatures almost immediately following denaturation.

Gene A sequence of DNA which codes for a polypeptide chain or, in a few cases, for ribosomal or transfer RNA.

Genome The entire repertoire of genes in a haploid or diploid cell or organism.

Genotype The genetic make-up of an individual, including traits not expressed in the phenotype since they are determined by recessive alleles in an individual who is heterozygous at these loci.

Germ cell A sperm or egg cell, or one of their precursor cells within the ovary or testes, being part of the germ line. Cells other than germ cells are referred to as somatic cells.

Heterogeneous nuclear RNA (hnRNA) The primary product of transcription in eukaryotes. It contains sequences complementary to both introns and exons in the gene coding strand. The intron sequences are cleaved out and the exon sequences spliced together to form messenger RNA, prior to the RNA moving from nucleus to cytoplasm.

Heterozygous Having different alleles at the two corresponding loci on homologous chromosomes.

Homologous chromosomes Chromosomes which pair up at meiosis, contain the same genetic loci, and are respectively derived from maternal and paternal origins.

Homozygous Having identical alleles at the two corresponding loci on homologous chromosomes.

Intron An intervening sequence in a eukaryote gene, represented in the hnRNA but not in the mRNA. Introns occur in most, but not all, eukaryotic genes and are spliced out of the hnRNA, allowing the remaining exon sequences to form the continuous sequence of the mRNA. Introns are not found in the genes of prokaryotes.

Linkage The phenomenon in which genes are located close to one another on the same chromosome and are therefore likely to segregate together. Sex-linked genes are on the sex chromosomes.

Locus The precise position on homologous chromosomes at which a specific gene is located.

Messenger RNA (mRNA) The RNA which carries the genetic message from the gene to the protein. mRNA is transcribed directly from DNA in prokaryotes, but in eukaryotes hnRNA is the primary transcription product and the mRNA is derived from hnRNA by exon splicing after intron removal. mRNA is translated into the polypeptide chains which form protein.

Mutation Any permanent change in the DNA; the smallest mutations are changes in single nucleotides, representing addition, loss or substitution, while large mutations may involve deletion, inversion or translocation of a substantial sequence. Many mutations do not persist because of DNA repair, and of those that do, only a few are phenotypically detectable. Most are deleterious but some may be fortuitously advantageous and spread by natural selection.

Oncogene Any gene which causes or helps to cause cancer. Oncogenes are carried by many retrovirus, and are similar or identical to sequences termed protooncogenes which occur in the human genome and code for substances such as growth factors or kinase enzymes.

Operon The DNA sequence in a bacterial genome which includes the coding genes and regulatory sequences for a series of coordinately regulated products. The best known operon is the 'lac' operon of *E. coli* which regulates synthesis of beta galactosidase enzyme and two other proteins involved in the lactose metabolic pathway.

Palindromic sequence A sequence which shows symmetry about a central axis point. Such sequences can adopt novel three-dimensional structures by base pairing between complementary parts of the same DNA strand. These structures may be important in protein recognition.

Phenotype Those aspects of the genetic make-up of an individual that are manifested in life. They represent the expressed part of the genotype, but do not include expression of recessive alleles present in a heterozygous condition.

Plasmid A factor consisting of a minicircle of DNA found in many bacteria and some eukaryotic cells. Plasmids may carry genes for drug resistance as well as sequences responsible for their own replication.

Processed pseudogene A sequence which is strikingly similar to another gene in the genome but lacks introns and is presumed to have arisen by reverse transcription of message. Pseudogenes are not transcribed since they contain many 'stop' codons.

Prokaryote An organism which has non-nucleated cells, in contrast to eukaryotes which have nucleated cells. All bacteria are prokaryotes while animals, plants and fungi are eukaryotes.

Promoter A regulatory region upstream from a gene coding sequence, with which the RNA polymerase molecule associates prior to transcription of the gene itself. Certain regions within the promoter show homology between many different genes. One such is the TATA box.

Pseudogene A gene which is not transcribed since it contains many 'stop' codons, but is similar to other functional genes in the genome.

Repetitious sequences DNA sequences which are repeated in the genome. Moderately repetitious sequences include genes for ribosomal and transfer RNA which are present as hundreds of near identical copies, while highly repetitious sequences are chiefly centromeric and contain thousands of very short sequences arranged in serial array.

Restriction enzyme An enzyme which cuts DNA at a precise sequence. They are found naturally in bacteria and are effective in degrading DNA of invading bacterio-phages. Such enzymes are now widely used in gene manipulation technology.

Retrovirus Any one of a group of small single-stranded RNA viruses that are characterised by their possessing reverse transcriptase enzyme, which transcribes the RNA genome into a DNA copy; this copy is then integrated into the host cell DNA. These viruses possess only 3 or 4 genes in their genome, the cancer-causing retroviruses having an oncogene sequence in addition to the other three genes. Non-oncogenic retroviruses include HIV (the causative agent of human AIDS) and visna virus which causes neurological disease in sheep. Both of these latter viruses belong to the subfamily of retroviruses called lentiviruses.

Ribosomal RNA (rRNA) A species of RNA which, together with specific ribosomal proteins, goes to form ribosomes. It is categorised according to its sedimentation constant, and that of eukaryotic cells is 28, 18, 5.8 and 5S. The coding sequences for these RNA species are present in the genome as multiple copies.

Semi-conservative replication The pattern followed in the replication of DNA, in which each original strand of the double helix is associated with a new daughter strand to yield the two double helical molecules that result from such replication.

Somatic cell Any cell that is not a germ cell, that is, other than a sperm cell or an egg or one of their cellular precursors.

Transfer RNA (tRNA) A species of RNA which is not translated into protein but functions as an amino acid carrier, transporting these molecules to ribosomes in the process of protein synthesis. A separate type of tRNA exists for each amino acid, each type having a codon triplet appropriate to that amino acid, permitting its precise location on the coding sequence of an mRNA molecule. Transfer RNA has a tertiary structure resembling a clover leaf and sediments at 4S.

Transgenic Of a cell or organism that has received and integrated into its genome one or more novel gene sequences usually following gene manipulation in the laboratory.

Transposon A sequence which can readily move about in the genome of an organism, inserting within or beside sequences with which it was not previously associated. Also called transposable elements, such sequences may be normal components of a particular genome, or, as in the case with retroviruses, essentially foreign to it. Transposons owe their mobility to their possession of specialised repeated sequences at either terminus of the linear molecule.

Upstream Running in a five prime direction with reference to the mRNA and therefore against the direction of transcription. Promoter sequences and initiation codons are upstream from the coding sequences which they help control.

Virion A single isolated virus particle.

Index

A

Acetabularia 61, 101, 102, 104
acetylation of histones 90–91
actin 21
AIDS virus 31
allele (allelmorph)
  definition of 129
allelic exclusion 113
allosteric protein 86
Alu family 23
annealing of DNA 7, 18
antibody 112
anti-codon 14, 16
anti-parallel strands of DNA 4
autoradiography 68, 106, 124
  definition of 129
autosome
  definition of 129
5 - azacytidine 92

B

Bacteria 25
bacterial genome 41
bacterial plasmid 45
bacteriophage 12
bacteriophage T4 32–34
bacteriophage lambda 35–39
banding patterns 70
base composition of DNA 118
Benzer, Seymour 2
beta globin gene 78, 92, 84, 85
beta thalassaemia 92

C

CAAT box 84
capping of mRNA 77
catabolic activator protein
(CAP) 86
  CAP site 86
CAT boxes 77
Cell
  commitment 99, 101, 105, 107–110
  cycle, definition of 129
  determination 99, 100, 108, 110
  differentiation 97–99, 108, 110
  lineage 111
  transdetermination 108, 112
  transdifferentiation 108, 109
centromere, definition of 129
chiasmata 61
chloroplast 25, 46, 47, 50
chloroplast DNA 48
chorion 112
chromatid, definition of 129
chromatin 25, 26, 52–60, 69, 72, 73,
    88–91, 98
  bands 68, 69
  definition of 129
chromocentre 66
chromomere 61, 63, 65, 69
  definition of 129
chromomeric DNA 63
  bands 89, 98
chromosome 1, 69, 112, 116
  composition 12
  dimunition 97
  function 6
  mini 27
  transcriptionally active 66
cistron 2
  definition of 129
clupeine 60
cohesive ends 37

col factor 46
colinearity 16
complementary DNA 106
 definition of 129
conjugation 44, 45
 definition of 129
constitutive heterochromatin 22, 71
cot curve 20, 21
CRO 38
 protein 38
crystallin 23, 105, 110
C Value 17
 paradox 18
cyclic AMP 86
cytochrome C 122

D

Degeneracy of genetic code 16, 118
 definition of 129
degenerate code 16
denaturation, definition of 129
depurination 117
DNase I 57, 89, 90, 98
DNA
 fingerprinting 124, 125
 glycosylase repair 117
 hybridisation 7
 methylation 91, 92, 95
 renaturation 20
 replication 7, 66, 73, 115
domains in protein 123
 definition of 129
downstream, definition of 129
Drosophila 17, 66, 68, 75, 92, 93, 95, 98–100, 112
 imaginal discs 109

E

E. coli 41–46, 51, 75, 80, 85, 87, 88, 92, 116
 genome 41
 lac operon 86
ELIZA system 105
enhancer 3
 definition of 130
 sequences 30, 84, 95
enzyme synthesis 75
erythroblast 108
erythrocyte 52, 106
 definition of 130
erythropoietic, definition of 130
 cascade 108
euchromatin 71
evolutionary relationships 122
exons 9, 10, 83, 84, 123
 definition of 130
 sequences of 82

F

Facultative heterochromatin 71
F factor 46
fibroin 68, 105
fibroin gene 113
flagellae 26
flanking sequence 69
foldback DNA 20–22
 definition of 130
frame shift mutation 41

G

Galactosidase enzyme 86
gene amplification 112
gene, definition of 130
gene regulatory protein 92, 105
gene splicing, V and C region 113
genetic code 4
genome 1
 definition of 130
genotype 1, 13
 definition of 130
germ cell, definition of 130
globin 2, 23, 68, 113, 120, 121
 promoter 84
globin messenger RNA 108
granulocyte 106
Griffith's experiment 12

H

Haemocoel 101
haemoglobin 23, 105, 119, 120, 123
haploid 17
 amounts of DNA 17
heat – shock genes 95
heterochromatin 71
heterozygous, definition of 130
highly repetitious DNA 20–22
 base composition of 22
 caesium chloride density 22
 centromeric regions of 22
histone 22, 25, 26, 52, 54, 56, 60, 71, 88,

91, 94, 105, 118, 120
histone acetylation 90, 95
histone Hl 90
HMG proteins 59, 91, 94
histone messenger RNA 74
hn RNA (heterogenous nuclear RNA) 72, 77, 83, 89, 124
  definition of 130
homeo-box genes 93, 95, 112
homeotic genes 93
homeotic mutants 112
homologue 66, 69
homologous chromosomes 124
  definition of 130
homozygous condition 112
  definition of 130
housekeeping genes 69
hydrogen bonds 4
hypersensitive sites 90

I

Icosahedral virus 30
induced enzyme sequences 85
interbands in polytene chromosome 68
interphase chromatin 71
interphase chromosomes 73
intron 3, 9, 10, 16, 17, 49, 50, 76, 77, 83, 84
  definition of 130
  sequences 82, 83
imaginal discs 99–101
immunoglobulin 105, 114
  synthesis 113
iridine 59
iris cells 109
IS elements 42

L

Lac operon 85, 86, 92
lagging strand of DNA 7
lampbrush chromosome 60–65
large ribosomal RNA genes 78, 79, 80
  molecules 112
leading strand of DNA 7
lens fibre cells 109
lenti virus 31
leptotene stage of meiotic chromosomes 67
linkage, definition of 130
linker DNA 56
locus 1
  definition of 130
lymphocyte 98, 105, 106
lysogenic cycle 38
lysogenic pathway 35
lysogeny 74
lysis 74
lytic pathway 35

M

Maturase system 49
megakaryocyte stem cell 106
melanin 109
Mendel, Gregor 1
Meselson and Stahl experiment 7
messenger RNA (mRNA) 3, 14, 40, 49, 64, 76, 82, 94, 105, 106
  definition of 131
metallothionein 112
metaphase chromosome 58, 70
methylation of DNA 6, 91, 92, 95
mitochondria 16, 25, 46–48, 50
mitochondrial DNA 47, 49, 126
mitochondrial genome 49
mitotic chromosome 69
mitotic banding of chromosome 69
moderately repetitious DNA 20–22
molecular reassociation 15
multipotential stem cell 106, 107
mutation 3, 12–14, 115, 116, 118
  definition of 131
  rates 18
muton 2
Mycoplasma 41, 44

N

Neutral mutation 120
nitrogen mustard 116
non-transcribed spacer 81
nuclear cage 73
nuclear lamina 73
nucleosome 25, 54, 56, 57, 59, 63, 64, 72, 73, 91, 120

O

Okazaki fragments 8
oncogenes 116
  evolution 126
  definition of 131
oncogenic viruses 31